Problem-Solving Guide with Solutions
to accompany Philip R. Kesten and David L. Tauck's

UNIVERSITY PHYSICS

for the Physical and Life Sciences   *Volume II*

Timothy A. French
DePaul University

W. H. FREEMAN AND COMPANY
New York

© 2013 by W. H. Freeman and Company
All rights reserved.
Printed in the United States of America

ISBN-13: 978-1-4641-0097-0
ISBN-10: 1-4641-0097-7

Second printing

W. H. Freeman and Company
41 Madison Avenue
New York, NY 10010
RG21 6XS England

www.whfreeman.com/physics

Contents

About the Author

DR. TIMOTHY FRENCH

Dr. Timothy French is currently a Visiting Assistant Professor in the Department of Chemistry at DePaul University in Chicago, IL. Dr. French holds a BS in chemistry from Rensselaer Polytechnic Institute and a MS and PhD in chemistry from Yale University. His undergraduate research looked at how to assess problem-solving ability in introductory physics students, while his doctoral research was on terahertz spectroscopy. Prior to arriving at DePaul, he was a Preceptor in Chemistry and Chemical Biology for four years at Harvard University where he taught an introductory physics sequence with biological and medical applications, experimental physical chemistry, and introductory physical chemistry, as well as the introductory physics class at the Harvard Summer School. He was awarded the Harold T. White Prize for Excellence in Teaching from the Harvard Physics Department in 2009.

A Note from the Author

The Problem-Solving Guide with Solutions is more than a book of answers to physics problems. It's designed to help you learn the way physicists approach and solve problems.

Many students, when approaching a physics problem, will half-read the question, pick out some of the variables used in the problem statement, and then start blindly hunting for an equation that shares the same letters, regardless of its utility.

Physicists, on the other hand, approach problems very differently. They'll fully read the question, underlining important information as they go. Next they will draw their own picture, even if one is provided, in order to quickly and easily organize and process the information provided. Then they will consider the underlying concept at play ("Is this a conservation of momentum question? Is Newton's second law best used here?") before determining which equation will be the most helpful. The equations are all manipulated algebraically before putting in numbers and arriving at a numerical answer.

Students usually move on to the next problem at this point without performing the crucial final step—checking the answer for reasonability. Physicists make sure the answer has the correct units, check some limiting cases ("What happens if the angle goes to 0 degrees? 90 degrees?"), and perform order of magnitude estimations all in order to gain confidence in their final answer.

Each solution in this book is set up to mirror this process with three distinct parts: **Set Up, Solve,** and **Reflect.** The **Set Up** portion contains all of the logic behind starting the problem, from a rehash of the important information in the problem statement to the conceptual underpinnings necessary in understanding the solution. The Solve portion contains the algebraic steps used to arrive at the numerical solution; note that this step comes *after* the Set Up step. Finally, the Reflect step provides a "sanity check" for the answer—"Is this what we expected? Does it make sense with respect to our observations of and interactions with everyday life? Does this answer seem reasonable?"

Very often instructors expect students to become expert problem solvers on their own without instruction. By explicitly showing the steps physicists use when solving problems, I hope you will be able to mirror and internalize these steps and make your foray into physics a more enjoyable and successful experience.

—Tim

Get Help with Premium Multimedia Resources

One of the benefits technology brings us in education is the ability to visualize concepts, gain problem-support, and test our skills outside of the traditional pen-and-paper classroom method. With that in mind, W. H. Freeman has developed a series of media assets geared at reinforcing conceptual understanding and building problem-solving skills.

P'Casts are videos that emulate the face-to-face experience of watching an instructor work a problem. Using a virtual whiteboard, the P'Casts' tutors demonstrate the steps involved in solving key worked examples, while explaining concepts along the way. The worked examples were chosen with the input of physics students and instructors across the country. P'Casts can be viewed online or downloaded to portable media devices.

Interactive Exercises are active learning, problem-solving activities. Each Interactive Exercise consists of a parent problem accompanied by a Socratic-dialog "help" sequence designed to encourage critical thinking as users do a guided conceptual analysis before attempting the mathematics. Immediate feedback for both correct and incorrect responses is provided through each problem-solving step.

Picture Its help bring static figures from the text to life. By manipulating variables within each animated figure students visualize a variety of physics concepts. Approximately 50 activities are available.

These Premium Multimedia Resources are available to you in a few places:

- If you're using the printed text, you can purchase the Premium Media Resources for a small fee via the Book Companion Website (BCS) at www.whfreeman.com/universityphysics1e.

- If you've purchased the W. H. Freeman media-enhanced eBook, these resources are embedded directly into the chapters of the text and available on the BCS.

- If you're using an online homework system, such as WebAssign or PhysicsPortal, these resources are integrated into your individual assignments and are available on the BCS.

And while all three of those places correlate the Premium Media Resources by chapter, we've gone one step further in the Problem-Solving Guide with Solutions and correlated each media resource by problem. Look for each relevant item called out after its corresponding problem:

    Get Help: Picture It - Adding and Subtracting Vectors

To get started using these exciting resources, log on to www.whfreeman.com/universityphysics1e!

Chapter 16
Electrostatics I

Conceptual Questions

16.5   The mass decreases because electrons are removed from the object to make it positively charged.

16.9   When the comb is run through your hair, electrons are transferred to the comb. The paper is polarized by the charged comb. The paper is then attracted to the comb. When they touch, a small amount of charge is transferred to the paper so now the paper and comb are similarly charged and repelled from each other.

16.13 Part a) If the charges creating the field move, the fact that the field propagates at the speed of light also allows us to understand the changes in the field, and hence force on the other charged objects. With the electric field, we see that charged particles take some time to experience the effects of other charges moving near them.

Part b) In electrostatics the field is just a computational device, and using it is merely a matter of convenience. However, in electrodynamics the field is necessary for energy and momentum to be conserved, so it is more than just convenience that leads us to the electric field.

16.17 Only charges inside the Gaussian surface contribute a nonzero flux, so the electric field of charges outside the surface can be ignored.

Multiple-Choice Questions

16.23 C (F). The forces must have the same magnitude due to Newton's 3rd law.
        Get Help: P'Cast 16.2 – Red Blood Cells

16.25 A (positively). Charge can move easily in a conductor, so the mobile electrons will be attracted to the positively charged rod, leaving the other end positively charged.

16.29 B ($-Q$). The electric field inside the conducting shell must be zero. The charge enclosed by a Gaussian sphere of radius r, where $R_1 < r < R_2$, must be zero. Therefore, the charge on the inner surface of the shell is $-Q$.

Estimation/Numerical Questions

16.33

$$F = \frac{kq_1q_2}{r^2}$$

$$q \sim \sqrt{\frac{(10\ \text{N})(10^{-1}\ \text{m})^2}{\left(10^{10}\dfrac{\text{N} \cdot \text{m}^2}{\text{C}^2}\right)}} = 10^{-5}\ \text{C}$$

Get Help: P'Cast 16.2 – Red Blood Cells

Problems

16.41

SET UP
We are asked to calculate the number of coulombs of negative charge in 0.5 kg of water. To answer this, we need to know the total number of electrons in the sample. The molar mass of water is 18 g/mol, and 1 mol of water contains 6.02×10^{23} molecules of water. Each molecule of water is made up of two hydrogen atoms and one oxygen atom. Hydrogen has one electron and oxygen has eight, which means each water molecule contains 10 electrons. The charge on one electron is -1.602×10^{19} C.

SOLVE

$$0.5\ \text{kg} \times \frac{1000\ \text{g}}{1\ \text{kg}} \times \frac{1\ \text{mol}}{18\ \text{g}} \times \frac{6.02 \times 10^{23}\ \text{molecules}}{1\ \text{mol}} \times \frac{10\ \text{electrons}}{1\ \text{molecule}} \times \frac{-1.602 \times 10^{-19}\ \text{C}}{1\ \text{electron}}$$

$$= \boxed{-2.7 \times 10^7\ \text{C}}$$

REFLECT
The net charge of 0.5 kg of water is zero since there are also 10 protons per molecule.

16.45

SET UP
We can use Coulomb's law to calculate the distance between two electrons such that the electrostatic force exerted by each on the other was equal in magnitude to the force of gravity on an electron. The mass of an electron is $m_e = 9.11 \times 10^{-31}$ kg; the magnitude of an electron's charge is $e = 1.602 \times 10^{-19}$ C.

SOLVE

$$\frac{k(e)(e)}{r^2} = m_e g$$

$$r = e\sqrt{\frac{k}{m_e g}} = (1.602 \times 10^{-19}\ \text{C})\sqrt{\frac{\left(8.99 \times 10^9 \dfrac{\text{N} \cdot \text{m}^2}{\text{C}^2}\right)}{(9.11 \times 10^{-31}\ \text{kg})\left(9.8\dfrac{\text{m}}{\text{s}^2}\right)}} = \boxed{5.08\ \text{m}}$$

REFLECT

The electric force between two electrons is repulsive, while the gravitational force between an electron and the Earth is attractive.

Get Help: P'Cast 16.2 – Red Blood Cells

16.49

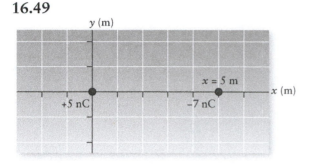

Figure 16-1 Problem 49

SET UP

Two charges, $q_A = +5$ nC and $q_B = -7$ nC, are located along the x-axis at $x = 0$ and $x = 5$ m, respectively. A third charge, $q_C = +2$ nC, is placed at a position x, where the net force acting on charge C is zero. To find this position, we can set the magnitude of the electric force of charge A on charge C equal to the magnitude of the electric force of charge B on charge C and solve for x.

SOLVE

$$F_{A \to C} = F_{B \to C}$$

$$\left| \frac{k q_A q_C}{x_{AC}^2} \right| = \left| \frac{k q_B q_C}{x_{BC}^2} \right|$$

$$\frac{|q_A|}{x^2} = \frac{|q_B|}{(x-5)^2} \text{ (assume distances are in m, charges in nC)}$$

$$(|q_A| - |q_B|)x^2 - 10|q_A|x + 25|q_A| = 0$$

$$x = \frac{10|q_A| \pm \sqrt{(10|q_A|)^2 - 4(|q_A| - |q_B|)(25|q_A|)}}{2(|q_A| - |q_B|)}$$

$$x = \frac{50 \pm \sqrt{2500 - 4(-2)(125)}}{2(-2)} = \frac{50 \pm \sqrt{3500}}{-4}$$

Taking the positive root:

$$x = \frac{50 + \sqrt{3500}}{-4} = \boxed{-27.3 \text{ m}}$$

REFLECT

The $(x - 5)^2$ in the denominator of $F_{B \to C}$ represents the fact that charge B is located 5 m to the right of the origin. It makes sense that charge C must be placed to the left of charge A since the magnitude of q_B is larger than the magnitude of q_A and $\vec{F}_{A \to C}$ is repulsive, while the $\vec{F}_{B \to C}$ is attractive.

Get Help: P'Cast 16.2 – Red Blood Cells

16.53

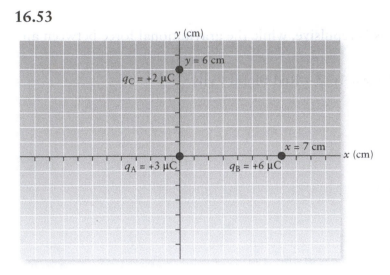

Figure 16-2 Problem 53

SET UP

Three positive charges ($q_A = 3 \times 10^{-6}$ C, $q_B = 6 \times 10^{-6}$ C, $q_C = 2 \times 10^{-6}$ C) are placed in a coordinate system (see figure). We can use Coulomb's law, $\vec{F} = \dfrac{kq_1 q_2}{r^2}\hat{r}$, to calculate the net force acting on each charge due to the two others. Since all of the charges are positive, all of the forces will be repulsive.

SOLVE

Charge A:

x component of net force on charge A:

$$\sum F_{q_A, x} = F_{q_B \to q_A, x} + F_{q_C \to q_A, x} = F_{q_B \to q_A, x} + 0 = -\frac{kq_B q_A}{r_{AB}^2}$$

$$= -\frac{\left(8.99 \times 10^9 \dfrac{\text{N} \cdot \text{m}^2}{\text{C}^2}\right)(6 \times 10^{-6}\ \text{C})(3 \times 10^{-6}\ \text{C})}{(7 \times 10^{-2}\ \text{m})^2} = -33.0\ \text{N}$$

y component of net force on charge A:

$$\sum F_{q_A, y} = F_{q_B \to q_A, y} + F_{q_C \to q_A, y} = 0 + F_{q_C \to q_A, y} = -\frac{kq_C q_A}{r_{AC}^2}$$

$$= -\frac{\left(8.99 \times 10^9 \dfrac{\text{N} \cdot \text{m}^2}{\text{C}^2}\right)(2 \times 10^{-6}\ \text{C})(3 \times 10^{-6}\ \text{C})}{(6 \times 10^{-2}\ \text{m})^2} = -15.0\ \text{N}$$

Magnitude of net force on charge A:

$$F_{q_A} = \sqrt{(-33.0\ \text{N})^2 + (-15.0\ \text{N})^2} = \boxed{36.3\ \text{N}}$$

Direction of net force on charge A:

$$\theta_A = \arctan\left(\frac{-15.0\ \text{N}}{-33.0\ \text{N}}\right) = \boxed{204.4°}\ \text{(which is 24.4° below the } -x\text{-axis)}.$$

Charge B:

x component of net force on charge B:

$$\sum F_{q_B,\, x} = F_{q_A \to q_B,\, x} + F_{q_C \to q_B,\, x} = \frac{kq_A q_B}{r_{AB}^2} + \frac{kq_C q_B}{r_{BC}^2}\cos(\theta_{C \to B})$$

$$= kq_B\left[\frac{q_A}{r_{AB}^2} + \frac{q_C}{r_{BC}^2}\left(\frac{7\ \text{cm}}{\sqrt{(6\ \text{cm})^2 + (7\ \text{cm})^2}}\right)\right]$$

$$= \left(8.99 \times 10^9 \frac{\text{N} \cdot \text{m}^2}{\text{C}^2}\right)(6 \times 10^{-6}\ \text{C})\left[\frac{3 \times 10^{-6}\ \text{C}}{(7 \times 10^{-2}\ \text{m})^2} + \frac{2 \times 10^{-6}\ \text{C}}{(6 \times 10^{-2}\ \text{m})^2 + (7 \times 10^{-2}\ \text{m})^2}(0.759)\right]$$

$$= 42.7\ \text{N}$$

y component of net force on charge B:

$$\sum F_{q_B,\, y} = F_{q_A \to q_B,\, y} + F_{q_C \to q_B,\, y} = 0 - \frac{kq_C q_B}{r_{BC}^2}\sin(\theta_{C \to B}) = -\frac{kq_B q_C}{r_{BC}^2}\left(\frac{6\ \text{cm}}{\sqrt{(6\ \text{cm})^2 + (7\ \text{cm})^2}}\right)$$

$$= -\frac{\left(8.99 \times 10^9 \frac{\text{N} \cdot \text{m}^2}{\text{C}^2}\right)(6 \times 10^{-6}\ \text{C})(2 \times 10^{-6}\ \text{C})}{(6 \times 10^{-2}\ \text{m})^2 + (7 \times 10^{-2}\ \text{m})^2}(0.651)$$

$$= -8.26\ \text{N}$$

Magnitude of net force on charge B:

$$F_{q_A} = \sqrt{(42.7\ \text{N})^2 + (-8.26\ \text{N})^2} = \boxed{43.5\ \text{N}}$$

Direction of net force on charge B:

$$\theta_A = \arctan\left(\frac{-8.26\ \text{N}}{42.7\ \text{N}}\right) = \boxed{349.1°}\ \text{(which is 10.9° below the } +x\text{-axis)}.$$

Charge C:

x component of net force on charge C:

$$\sum F_{q_C,\, x} = F_{q_A \to q_C,\, x} + F_{q_B \to q_C,\, x} = 0 - \frac{kq_B q_C}{r_{BC}^2}\cos(\theta_{B \to C}) = -\frac{kq_B q_C}{r_{BC}^2}\left(\frac{7\ \text{cm}}{\sqrt{(6\ \text{cm})^2 + (7\ \text{cm})^2}}\right)$$

$$= -\frac{\left(8.99 \times 10^9 \frac{\text{N} \cdot \text{m}^2}{\text{C}^2}\right)(6 \times 10^{-6}\ \text{C})(2 \times 10^{-6}\ \text{C})}{(6 \times 10^{-2}\ \text{m})^2 + (7 \times 10^{-2}\ \text{m})^2}(0.759)$$

$$= -9.63\ \text{N}$$

y component of net force on charge C:

$$\sum F_{q_C, y} = F_{q_A \to q_C, y} + F_{q_B \to q_C, y} = \frac{kq_A q_C}{r_{AC}^2} + \frac{kq_B q_C}{r_{BC}^2} \sin(\theta_{B \to C}) = kq_C \left[\frac{q_A}{r_{AC}^2} + \frac{q_B}{r_{BC}^2} \left(\frac{6 \text{ cm}}{\sqrt{(6 \text{ cm})^2 + (7 \text{ cm})^2}} \right) \right]$$

$$= \left(8.99 \times 10^9 \frac{\text{N} \cdot \text{m}^2}{\text{C}^2} \right) (2 \times 10^{-6} \text{ C}) \left[\frac{(3 \times 10^{-6} \text{ C})}{(6 \times 10^{-2} \text{ m})^2} + \frac{(6 \times 10^{-6} \text{ C})}{(6 \times 10^{-2} \text{ m})^2 + (7 \times 10^{-2} \text{ m})^2} (0.651) \right]$$

$$= 23.2 \text{ N}$$

Magnitude of net force on charge C:

$$F_{q_A} = \sqrt{(-9.63 \text{ N})^2 + (23.2 \text{ N})^2} = \boxed{25.2 \text{ N}}$$

Direction of net force on charge C:

$$\theta_A = \arctan\left(\frac{23.2 \text{ N}}{-9.63 \text{ N}} \right) = \boxed{112.6°} \text{ (which is } 67.4° \text{ above the } -x\text{-axis).}$$

REFLECT
We could have also represented all of the forces in terms of unit vectors:
$$\sum \vec{F}_{q_A} = -(33.0 \text{ N})\hat{x} - (15.0 \text{ N})\hat{y}, \sum \vec{F}_{q_B} = (42.7 \text{ N})\hat{x} - (8.26 \text{ N})\hat{y}, \text{ and } \sum \vec{F}_{q_C} = -(9.63 \text{ N})\hat{x} + (23.2 \text{ N})\hat{y}.$$

Get Help: P'Cast 16.2 – Red Blood Cells

16.57

SET UP
A person walking across a carpet accumulates -50 nC of charge in each step. Multiplying this value by 25 will tell us the amount of charge she builds up in 25 steps. Dividing the total charge on the person by the charge on a single electron (-1.602×10^{-19} C) yields the total number of excess electrons present due to static buildup. Finally, from these same values, we can calculate the number of steps required to accumulate a total of 10^{12} electrons.

SOLVE

Part a)

$$25 \text{ steps} \times \frac{-50 \text{ nC}}{1 \text{ step}} = \boxed{-1250 \text{ nC} = -1.25 \ \mu\text{C}}$$

Part b)

$$-1.25 \times 10^{-6} \text{ C} \times \frac{1 \text{ electron}}{-1.602 \times 10^{-19} \text{ C}} = \boxed{7.80 \times 10^{12} \text{ electrons}}$$

Part c)

$$10^{12} \text{ electrons} \times \frac{-1.602 \times 10^{-19} \text{ C}}{1 \text{ electron}} \times \frac{1 \text{ step}}{-50 \times 10^{-9} \text{ C}} = 3.2 \text{ steps}$$

A worker $\boxed{\text{should not take more than 3 steps}}$ before touching the components.

REFLECT

People building electronics or working with precise instrumentation will "ground" themselves often by touching a metal conduit or wearing an antistatic wrist strap to rid themselves of excess charge buildup.

16.67

SET UP

A rod of length $L = 1.0$ m has a positive charge of $Q = 1.0$ C evenly distributed throughout. We are asked to calculate the electric field strength at a point P that is a distance $y_P = 0.50$ m from the middle of the rod along a line bisecting the rod (see figure). We'll set up a coordinate system where the origin is at the center of the rod with $+x$ pointing to the right and $+y$ pointing up towards P. Due to symmetry, the x component of the electric field at point P is equal to zero. Therefore, the electric field strength at P is equal to the magnitude of the y component of the electric field, E_y. We can split the rod up into infinitesimal point charges dq and integrate over the y components of the field due to each point charge, $dE_y = (dE)\cos(\theta)$ where θ is the angle made with the y-axis. The field dE due to the point charge dq is equal to $dE = \dfrac{kdq}{r^2}$, where r is the straight-line distance between dq and P. We can convert the integral from dq to dx by realizing the charge on the rod is uniformly distributed (*i.e.*, the linear charge density is constant).

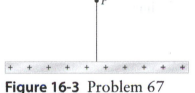

Figure 16-3 Problem 67

SOLVE

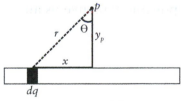

Figure 16-4 Problem 67

$$E_y = \int dE_y = \int (dE)\cos(\theta) = \int \left(\frac{kdq}{r^2}\right)\left(\frac{y_P}{r}\right)$$

Converting the integral from dq to dx using $\dfrac{dq}{Q} = \dfrac{dx}{L}$:

$$E_y = \int_{-\frac{L}{2}}^{\frac{L}{2}} \left(\frac{k}{r^2}\right)\left(\frac{y_P}{r}\right)\left(\frac{Q}{L}dx\right) = \frac{kQy_P}{L}\int_{-\frac{L}{2}}^{\frac{L}{2}} \frac{dx}{r^3}$$

But $r = \sqrt{x^2 + y_P^2} = (x^2 + y_P^2)^{\frac{1}{2}}$:

$$E_y = \frac{kQy_P}{L}\int_{-\frac{L}{2}}^{\frac{L}{2}} \frac{dx}{(x^2 + y_P^2)^{\frac{3}{2}}} = \frac{kQy_P}{L}\left[\frac{x}{y_P^2\sqrt{x^2 + y_P^2}}\right]_{-\frac{L}{2}}^{\frac{L}{2}} = \frac{kQ}{Ly_P}\left[\frac{\left(\frac{L}{2}\right)}{\sqrt{\left(\frac{L}{2}\right)^2 + y_P^2}} - \frac{\left(-\frac{L}{2}\right)}{\sqrt{\left(-\frac{L}{2}\right)^2 + y_P^2}}\right]$$

$$= \frac{kQ}{Ly_P}\left[\frac{L}{\sqrt{\left(\frac{L}{2}\right)^2 + y_P^2}}\right] = \frac{kQ}{y_P}\left[\frac{1}{\sqrt{\left(\frac{L}{2}\right)^2 + y_P^2}}\right]$$

$$= \frac{\left(8.99 \times 10^9 \frac{\text{N} \cdot \text{m}^2}{\text{C}^2}\right)(1.0 \text{ C})}{0.50 \text{ m}}\left[\frac{1}{\sqrt{\left(\frac{1.0 \text{ m}}{2}\right)^2 + (0.50 \text{ m})^2}}\right] = \boxed{2.5 \times 10^{10}\frac{\text{N}}{\text{C}}}$$

REFLECT

As L approaches infinity, our algebraic answer is equal to the electric field due to a very long, straight wire, $E = \dfrac{\lambda}{2\pi\epsilon_0 r}$.

Get Help: Interactive Example – Continuous Line of Charge

16.71

SET UP

The electric field at a point along the central axis of a ring of charge Q is $E_y(y) = \dfrac{kQy}{(R^2 + y^2)^{\frac{3}{2}}}$.
In order to find the position where the field is a maximum, we need to set the derivative of E_y with respect to y equal to zero and solve for y. Evaluating E_y at this position will give us an expression for the maximum electric field.

SOLVE

Part a)

$$\frac{dE_y}{dy} = 0$$

$$\frac{d}{dy}\left[\frac{kQy}{(R^2 + y^2)^{\frac{3}{2}}}\right] = kQ\frac{d}{dy}[y(R^2 + y^2)^{-\frac{3}{2}}] = 0$$

$$\left[(R^2 + y^2)^{-\frac{3}{2}} - \frac{3}{2}y(2y)\right] = (R^2 + y^2)^{-\frac{5}{2}}[(R^2 + y^2) - 3y^2] = (R^2 + y^2)^{-\frac{5}{2}}[R^2 - 2y^2] = 0$$

$$y = \sqrt{\frac{R^2}{2}} = \boxed{\frac{R}{\sqrt{2}}}$$

Part b)

$$E_y\left(y = \frac{R}{\sqrt{2}}\right) = \frac{kQ\left(\dfrac{R}{\sqrt{2}}\right)}{\left(R^2 + \left(\dfrac{R}{\sqrt{2}}\right)^2\right)^{\frac{3}{2}}} = \frac{kQR}{\sqrt{2}\left(\dfrac{3R^2}{2}\right)^{\frac{3}{2}}} = \boxed{\frac{2kQ}{R^2\sqrt{27}}}$$

REFLECT

The magnitude of the field will be the same for an equal distance above and below the ring by symmetry.

Get Help: Interactive Example – Continuous Line of Charge

16.75

SET UP

A uniformly charged plastic rod ($L = 0.100$ m) is sealed inside of a plastic bag. The net electric flux through the bag is $\Phi = 7.5 \times 10^5 \dfrac{\text{N} \cdot \text{m}^2}{\text{C}}$. We can use Gauss' law, $\Phi = \dfrac{q_{\text{encl}}}{\epsilon_0}$, to calculate the linear charge density of the rod.

SOLVE

$$\Phi = \frac{q_{\text{encl}}}{\epsilon_0} = \frac{\lambda L}{\epsilon_0}$$

$$\lambda = \frac{\Phi \epsilon_0}{L} = \frac{\left(7.5 \times 10^5 \dfrac{\text{N} \cdot \text{m}^2}{\text{C}}\right)\left(8.85 \times 10^{-12} \dfrac{\text{C}^2}{\text{N} \cdot \text{m}^2}\right)}{0.100 \text{ m}} = \boxed{6.6 \times 10^{-5} \dfrac{\text{C}}{\text{m}}}$$

REFLECT

The net electric flux through the bag is positive, which means the rod is positively charged. The exact shape of the bag is irrelevant because we are given the flux.

16.79

SET UP

A sphere carries a uniform surface charge density (charge per unit area) σ. We can use Gauss' law to find an expression for the electric field just outside the surface of the sphere. At this location, the Gaussian sphere will have approximately the same surface area as the charged sphere, which we'll call A. Assuming σ is positive, the electric field will point radially outward from the sphere.

SOLVE

$$\oint \vec{E} \cdot d\vec{A} = \frac{q_{\text{encl}}}{\epsilon_0}$$

$$EA = \frac{(\sigma A)}{\epsilon_0}$$

$$\boxed{E = \frac{\sigma}{\epsilon_0}, \text{ pointing radially outward}}$$

REFLECT
The electric field is constant just outside the sphere.

16.81

SET UP
A very long, hollow, charge cylinder has an inner radius $a = 3.00 \times 10^{-2}$ m, an outer radius $b = 5.00 \times 10^{-2}$ m, and a uniform charge density $\rho = +42.0 \times 10^{-6}$ C/m³. Since the charge distribution has cylindrical symmetry, we will use a cylindrical Gaussian surface of (constant) length L and (variable) radius r, split the problem into three regions—1) $r \leq a$, 2) $a \leq r \leq b$, and 3) $r \geq b$—and apply Gauss' law in order to find the electric field in each region. The charge enclosed by the Gaussian cylinder in region 1 is zero, which means the electric field in that region is also zero. The cylinder in region 2 encloses a fraction of the charge distributed throughout the thick, cylindrical shell; the volume of this shape can be calculated by subtracting the volume of the cylindrical hole from the volume of the Gaussian cylinder. Finally, the Gaussian cylinder in region 3 encloses the entire charged cylinder.

SOLVE

Part a)

$r \leq a$:

Looking down the cylinder:

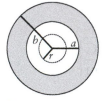

Figure 16-5 Problem 81

$$\oint \vec{E} \cdot d\vec{A} = \frac{q_{encl}}{\epsilon_0}$$

$q_{encl} = 0$; therefore, $\boxed{\vec{E} = 0}$

Part b)

$a \leq r \leq b$:

Looking down the cylinder:

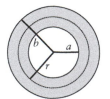

Figure 16-6 Problem 81

$$\oint \vec{E} \cdot d\vec{A} = \frac{q_{encl}}{\epsilon_0}$$

$$EA = \frac{(\rho V_{encl})}{\epsilon_0} = \frac{\rho(\pi r^2 - \pi a^2)L}{\epsilon_0}$$

$$E(2\pi rL) = \frac{\rho(\pi r^2 - \pi a^2)L}{\epsilon_0}$$

$$E = \frac{\rho(r^2 - a^2)}{2\epsilon_0 r} = \frac{\left(42.0 \times 10^{-6}\frac{C}{m^3}\right)(r^2 - a^2)}{2\left(8.85 \times 10^{-12}\frac{C^2}{N \cdot m^2}\right)r} = \left(2.37 \times 10^6\frac{N}{C \cdot m}\right)\frac{(r^2 - a^2)}{r}$$

The magnitude of the electric field in the region $a \leq r \leq b$ has a magnitude of

$$\boxed{\left(2.37 \times 10^6\frac{N}{C \cdot m}\right)\frac{(r^2 - a^2)}{r} \text{ and points radially outward}}.$$

Part c)

$r \geq b$:

Looking down the cylinder:

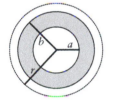

Figure 16-7 Problem 81

$$\oint \vec{E} \cdot d\vec{A} = \frac{q_{encl}}{\epsilon_0}$$

$$EA = \frac{(\rho V)}{\epsilon_0} = \frac{\rho(\pi b^2 - \pi a^2)L}{\epsilon_0}$$

$$E(2\pi rL) = \frac{\rho(\pi b^2 - \pi a^2)L}{\epsilon_0}$$

$$E = \frac{\rho(b^2 - a^2)}{2\epsilon_0 r} = \frac{\left(42.0 \times 10^{-6}\frac{C}{m^3}\right)((5.00 \times 10^{-2}\ m)^2 - (3.00 \times 10^{-2}\ m)^2)}{2\left(8.85 \times 10^{-12}\frac{C^2}{N \cdot m^2}\right)r} = \frac{\left(3800\frac{N \cdot m}{C}\right)}{r}$$

The magnitude of the electric field in the region $r \geq b$ has a

$$\boxed{\text{magnitude of } \frac{\left(3800\frac{N \cdot m}{C}\right)}{r} \text{ and points radially outward}}.$$

REFLECT

Problems that ask you to "determine the electric field for all radii" or "determine the electric field everywhere in space" will usually require you to invoke Gauss' law. The field should increase as r increases in region 2 since the Gaussian surface is enclosing more and more charge. The field should decrease as r increases in region 3 since we are moving farther and farther away from the source of charge.

Get Help: Interactive Example – Coaxial Cylindrical Conductors
Interactive Example – Spherical Shell Insulator

16.85

SET UP

A long, straight rod has a linear charge density of $\lambda = 12 \times 10^{-6}\dfrac{C}{m}$. We can use Gauss' law to calculate the magnitude of the electric field at point P, which is a distance $r = 0.10$ m radially out from the central axis of the rod. We will use a cylindrical Gaussian surface of radius r and length L. The charge enclosed by this surface is equal to $q_{encl} = \lambda L$. The electric field points radially outward from the rod everywhere in space because the linear charge density is positive; we can represent this using $\hat{r}$.

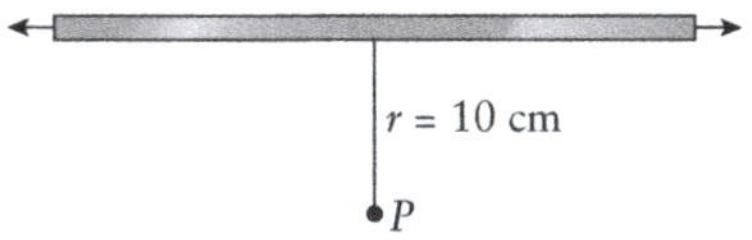

Figure 16-8 Problem 85

SOLVE

Magnitude of the electric field at P:

$$\oint \vec{E} \cdot d\vec{A} = \frac{q_{encl}}{\epsilon_0}$$

$$E(2\pi r L) = \frac{(\lambda L)}{\epsilon_0}$$

$$E = \frac{\lambda}{2\pi\epsilon_0 r} = \frac{\left(12 \times 10^{-6}\dfrac{C}{m}\right)}{2\pi\left(8.85 \times 10^{-12}\dfrac{C^2}{N \cdot m^2}\right)(0.10 \text{ m})} = 2.2 \times 10^6 \frac{N}{C}$$

Electric field vector:

$$\boxed{\vec{E} = \left(2.2 \times 10^6\frac{N}{C}\right)\hat{r}}$$

REFLECT

Your exact answer depends on how you've defined your coordinate system. For example, if we defined a Cartesian coordinate system in the drawing where point P lies along the $-y$-axis, we can represent the electric field at point P as $\vec{E} = -\left(2.2 \times 10^6\frac{N}{C}\right)\hat{y}$. Although convenient for point P, this coordinate system would not be as useful for describing the electric field at points that lie off axis.

16.89

SET UP

A red blood cell carries an excess charge of $Q = -2.5 \times 10^{-12}$ C distributed uniformly over its surface. We will model the cells as spheres of diameter $d = 7.5 \times 10^{-6}$ m and mass $m = 9.0 \times 10^{-14}$ kg. The number of excess electrons on the red blood cell is equal to Q divided by the charge on one electron, -1.602×10^{-19} C. In order to determine whether these extra electrons appreciably affect the mass of the cell, we can calculate the ratio of the mass of the excess electrons to the mass of the sphere; the mass of an electron is $m_e = 9.11 \times 10^{-31}$ kg. Finally, the surface charge density of the red blood cell is equal to the total charge divided by the surface area of the cell.

SOLVE

Part a)

$$-2.5 \times 10^{-12} \text{ C} \times \frac{1 \text{ electron}}{-1.602 \times 10^{-19} \text{ C}} = \boxed{1.6 \times 10^7 \text{ electrons}}$$

Part b)

$$\frac{m_{\text{electrons}}}{m_{\text{cell}}} = \frac{(1.6 \times 10^7)(9.11 \times 10^{-31} \text{ kg})}{9.0 \times 10^{-14} \text{ kg}} = \boxed{1.6 \times 10^{-10}}$$

The extra mass is $\boxed{\text{not significant}}$.

Part c)

$$\sigma = \frac{Q}{A} = \frac{Q}{4\pi R^2} = \frac{Q}{4\pi \left(\dfrac{d}{2}\right)^2} = \frac{Q}{\pi d^2} = \frac{-2.5 \times 10^{-12} \text{ C}}{\pi (7.5 \times 10^{-6} \text{ m})^2}$$

$$= \boxed{-1.4 \times 10^{-2} \frac{\text{C}}{\text{m}^2}} \times \frac{1 \text{ electron}}{-1.6 \times 10^{-19} \text{ C}} = \boxed{8.8 \times 10^{16} \frac{\text{electrons}}{\text{m}^2}}$$

REFLECT

There would need to be about 10^{17} excess electrons, or an excess charge of around 10^{-2} C, for the mass of the excess electrons to be approximately equal to the mass of the cell.

16.93

SET UP

The elephant nose fish can detect changes in an electric field as small as 3.0×10^{-6} N/C. We can set this equal to the expression for the electric field due to a point charge, $E = \dfrac{kQ}{r^2}$, in order to calculate the minimum charge required to create a field of this strength at a distance of $r = 0.75$ m. This charge divided by the magnitude of the charge on one electron (1.602×10^{-19} C) is equal to the number of electrons required to achieve this charge.

SOLVE

Part a)

$$E = \frac{kQ}{r^2}$$

$$Q = \frac{Er^2}{k} = \frac{\left(3.0 \times 10^{-6}\dfrac{\text{N}}{\text{C}}\right)(0.75 \text{ m})^2}{\left(8.99 \times 10^9\dfrac{\text{N} \cdot \text{m}^2}{\text{C}^2}\right)} = \boxed{1.9 \times 10^{-16} \text{ C}}$$

Part b)

$$1.9 \times 10^{-16} \text{ C} \times \frac{1 \text{ electron}}{1.602 \times 10^{-19} \text{ C}} = \boxed{1200 \text{ electrons}}$$

REFLECT

The fish can only detect *changes* in the local electric field of 3.0×10^{-6} N/C, so the additional field can either add to or subtract from the unperturbed field.

> **Get Help:** Picture It – Electric Field
> Interactive Example – Zero of E field Two Charges
> P'Cast 16.4 – Electric Field Is Zero

16.97

SET UP

Six positive charges ($q_1 = 1 \times 10^{-3}$ C, $q_2 = 2 \times 10^{-3}$ C, $q_3 = 3 \times 10^{-3}$ C, $q_4 = 4 \times 10^{-3}$ C, $q_5 = 5 \times 10^{-3}$ C, $q_6 = 6 \times 10^{-3}$ C) are arranged in a regular hexagon with 5-cm-long sides (see figure). The electric field at the center of the hexagon is equal to the vector sum of the electric fields due to each point charge. The magnitude of the electric field due to a point charge q at a distance r away is $E = \dfrac{kq}{r^2}$. Since all of the charges are positive, the electric field will point away from each charge towards the center of the hexagon.

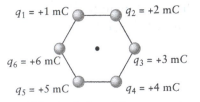

$q_1 = +1$ mC $q_2 = +2$ mC
$q_6 = +6$ mC $q_3 = +3$ mC
$q_5 = +5$ mC $q_4 = +4$ mC

Figure 16-9 Problem 97

SOLVE

Magnitudes of the field due to each point charge at the center:

$$E_1 = \frac{kq_1}{r_1^2} = \frac{\left(8.99 \times 10^9\dfrac{\text{N} \cdot \text{m}^2}{\text{C}^2}\right)(1 \times 10^{-3} \text{ C})}{(5 \times 10^{-2} \text{ m})^2} = 3.596 \times 10^9\frac{\text{N}}{\text{C}}$$

$$E_2 = \frac{kq_2}{r_2^2} = \frac{k(2q_1)}{r_1^2} = 2E_1$$

$$E_3 = \frac{kq_3}{r_3^2} = \frac{k(3q_1)}{r_1^2} = 3E_1$$

$$E_4 = \frac{kq_4}{r_4^2} = \frac{k(4q_1)}{r_1^2} = 4E_1$$

$$E_5 = \frac{kq_5}{r_5^2} = \frac{k(5q_1)}{r_1^2} = 5E_1$$

$$E_6 = \frac{kq_6}{r_6^2} = \frac{k(6q_1)}{r_1^2} = 6E_1$$

x component of the field at the center:

$$E_x = E_1\cos(60°) - E_2\cos(60°) - E_3 - E_4\cos(60°) + E_5\cos(60°) + E_6$$

$$= [E_1 - E_2 - E_4 + E_5]\cos(60°) + E_6 - E_3 = [E_1 - 2E_1 - 4E_1 + 5E_1]\cos(60°) + 6E_1 - 3E_1$$

$$= [1 - 2 - 4 + 5]E_1\cos(60°) + 6E_1 - 3E_1 = 3E_1 = 3\left(3.596 \times 10^9\frac{\text{N}}{\text{C}}\right) = 1.079 \times 10^{10}\frac{\text{N}}{\text{C}}$$

y component of the field at the center:

$$E_y = -E_1\sin(60°) - E_2\sin(60°) + E_4\sin(60°) + E_5\sin(60°)$$

$$= [-E_1 - E_2 + E_4 + E_5]\sin(60°) = [-E_1 - 2E_1 + 4E_1 + 5E_1]\sin(60°)$$

$$= [-1 - 2 + 4 + 5]E_1\sin(60°) = \frac{6\sqrt{3}}{2}E_1 = 3\sqrt{3}\left(3.596 \times 10^9\frac{\text{N}}{\text{C}}\right) = 1.869 \times 10^{10}\frac{\text{N}}{\text{C}}$$

Electric field at the center has a magnitude of $2.16 \times 10^{10}\frac{\text{N}}{\text{C}}$ with components of $1.08 \times 10^{10}\frac{\text{N}}{\text{C}}$ to the right and $1.87 \times 10^{10}\frac{\text{N}}{\text{C}}$ above the horizontal .

REFLECT

There is more positive charge to the left of the center than the right and below than above, so the electric field at the center should point up and to the right. The field must point at an angle of 60 degrees above the horizontal due to symmetry.

16.101

SET UP

In one model, parts of a cloud can be viewed as two round, parallel, oppositely charged sheets with a radius of $R = 2.5 \times 10^3$ m and a distance of 6.0×10^3 m apart. The field midway between the two sheets has a magnitude of 2.0×10^5 N/C. Using the expression for the magnitude of the electric field due to a round disk, $E_{\text{disk}} = \left(\frac{2kQ}{R^2}\right)\left(1 - \frac{y}{\sqrt{R^2 + y^2}}\right)$, we can calculate the magnitude of the charge Q. Another model treats the parts of the cloud as equal but opposite point charges. Again, the field midway between the two charges has a magnitude of 2.0×10^5 N/C. We can use the expression for the magnitude of the electric field due to a point charge, $E_{\text{point charge}} = \dfrac{kQ}{R^2}$, to calculate Q in this case.

SOLVE
Part a)

$$\vec{E}_{\text{total}} = \vec{E}_1 + \vec{E}_2 = E_{\text{disk}}(-\hat{y}) + E_{\text{disk}}(-\hat{y}) = -2E_{\text{disk}}\hat{y}$$

$$E_{\text{total}} = 2E_{\text{disk}} = 2\left(\frac{2kQ}{R^2}\right)\left(1 - \frac{y}{\sqrt{R^2 + y^2}}\right)$$

$$Q = \frac{E_{\text{total}}R^2}{4k\left(1 - \dfrac{y}{\sqrt{R^2 + y^2}}\right)}$$

$$= \frac{\left(2.0 \times 10^5 \dfrac{\text{N}}{\text{C}}\right)(2.5 \times 10^3 \text{ m})^2}{4\left(8.99 \times 10^9 \dfrac{\text{C}^2}{\text{N} \cdot \text{m}^2}\right)\left(1 - \dfrac{3.0 \times 10^3 \text{ m}}{\sqrt{(2.5 \times 10^3 \text{ m})^2 + (3.0 \times 10^3 \text{ m})^2}}\right)} = \boxed{150 \text{ C}}$$

Part b)

$$E_{\text{total}} = 2E_{\text{point charge}} = 2\left(\frac{kQ}{R^2}\right)$$

$$Q = \frac{E_{\text{total}}R^2}{2k} = \frac{\left(2.0 \times 10^5 \dfrac{\text{N}}{\text{C}}\right)(3.0 \times 10^3 \text{ m})^2}{2\left(8.99 \times 10^9 \dfrac{\text{C}^2}{\text{N} \cdot \text{m}^2}\right)} = \boxed{100 \text{ C}}$$

REFLECT
Even though the two models are very simplistic, their results agree rather well. Actual measured charges are on the order of a few tens of coulombs.

16.105

SET UP
The nucleus of an iron atom contains 26 protons and has a radius of $R_{\text{nuc}} = 4.6 \times 10^{-15}$ m. The radius of the atom is 0.50×10^{-10} m. We can treat the nucleus as a positive point charge of magnitude $26e$ for all points outside of the nucleus. The magnitude of the electric field due to a point charge is $E = \dfrac{kq}{r^2}$; the field will point radially outward because the nucleus is positively charged. We'll assume that the net force acting on the outermost electron is equal to the force due to the electric field of the nucleus at that point. We can calculate that electron's acceleration using Newton's second law.

SOLVE

Part a)

$$E = \frac{kq}{r^2} = \frac{k(26e)}{r_{\text{nuc}}^2} = \frac{\left(8.99 \times 10^9 \frac{\text{N} \cdot \text{m}^2}{\text{C}^2}\right)(26)(1.6 \times 10^{-19} \text{ C})}{(4.6 \times 10^{-15} \text{ m})^2} = 1.8 \times 10^{21} \frac{\text{N}}{\text{C}}$$

The electric field has a magnitude of $\boxed{1.8 \times 10^{21} \frac{\text{N}}{\text{C}} \text{ and points radially outward}}$.

Part b)

$$E = \frac{kq}{r^2} = \frac{k(26e)}{r_{\text{atom}}^2} = \frac{\left(8.99 \times 10^9 \frac{\text{N} \cdot \text{m}^2}{\text{C}^2}\right)(26)(1.6 \times 10^{-19} \text{ C})}{(0.5 \times 10^{-10} \text{ m})^2} = 1.5 \times 10^{13} \frac{\text{N}}{\text{C}}$$

The electric field has a magnitude of $\boxed{1.5 \times 10^{13} \frac{\text{N}}{\text{C}} \text{ and points radially outward}}$.

Part c)

$$\sum F = F_{\text{Coul}} = |q|E = m_e a$$

$$a = \frac{|q|E}{m_e} = \frac{(1.6 \times 10^{-19} \text{ C})\left(1.5 \times 10^{13} \frac{\text{N}}{\text{C}}\right)}{9.11 \times 10^{-31} \text{ kg}} = 2.6 \times 10^{24} \frac{\text{m}}{\text{s}^2}$$

The acceleration of the outermost electron has a magnitude of $\boxed{2.6 \times 10^{24} \frac{\text{m}}{\text{s}^2} \text{ and}}$ $\boxed{\text{points radially inward}}$.

REFLECT

The force acting on the outermost electron will be much less than this due to the effects of the other 25 electrons. These electrons are closer to the nucleus and "screen" the charge of the nucleus so the effective field at the position of the outermost electron is smaller.

16.109

SET UP

A sphere of radius R has uniform charge density ρ. A closed Gaussian surface consisting of a circular disk and a hemisphere, each of radius r and concentric with the sphere, surrounds half of the sphere as shown in the figure. The electric field emanates radially outward from the charged sphere as if it were a point charge. Therefore, the flux through the disk is equal to 0. The flux through the hemisphere is equal to EA, where E is the expression for the magnitude of the electric field and A is the surface area of the hemisphere. We can rearrange our expression to show that the total flux through the Gaussian surface is equal to the total enclosed charge divided by ε_0; only the charge located in the top half of the sphere is enclosed by the Gaussian surface.

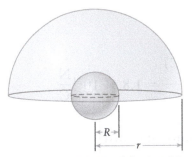

Figure 16-10 Problem 109

SOLVE

Part a)

The | flux through the circular disk is zero | because the electric field vectors in that plane are perpendicular to the normal vector of the disk.

Flux through the hemisphere:

$$\Phi = \int \vec{E} \cdot d\vec{A} = EA = \left(\frac{kQ}{r^2}\right)(2\pi r^2) = 2\pi k(\rho V) = 2\pi k\rho\left(\frac{4}{3}\pi R^3\right) = \frac{8\pi^2 k\rho R^3}{3}$$

$$= \frac{8\pi^2 \rho R^3}{3}\left(\frac{1}{4\pi\varepsilon_0}\right) = \boxed{\frac{2\pi\rho R^3}{3\varepsilon_0}}$$

Part b)

$$\Phi_{total} = \frac{2\pi\rho R^3}{3\varepsilon_0} = \frac{\left(\frac{2\pi\rho R^3}{3}\right)}{\varepsilon_0} = \frac{\rho\left(\frac{2\pi R^3}{3}\right)}{\varepsilon_0} = \frac{\rho\left(\frac{V}{2}\right)}{\varepsilon_0} = \frac{q_{encl}}{\varepsilon_0}$$

REFLECT

Gauss' law holds for any closed Gaussian surface.

16.113

SET UP

Two hollow, concentric, spherical shells are covered with charge. The inner sphere has a radius R_i and a surface charge density of $+\sigma_i$; the outer sphere has a radius R_o and a surface charge density of $-\sigma_o$. We can use Gauss' law to derive an expression for the electric field everywhere in space. We will use a spherical Gaussian surface of radius r because of the symmetry of the charge distribution. Since the spheres are hollow, the charge only exists on the surface of each sphere. Once we have an algebraic expression for the electric field, we can plug in values $R_i = 5$ cm, $R_o = 8$ cm, $\sigma_i = 20 \times 10^{-6} \dfrac{C}{cm^2}$, and $\sigma_o = 14 \times 10^{-6} \dfrac{C}{cm^2}$ to calculate the magnitude of the field in each region.

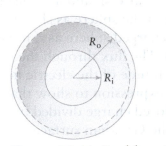

Figure 16-11 Problem 113

SOLVE

Part a)

$r < R_i$

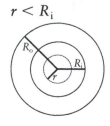

$$\int \vec{E} \cdot d\vec{A} = \frac{q_{encl}}{\varepsilon_0} = \frac{0}{\varepsilon_0}$$

$$\boxed{\vec{E} = 0}$$

Part b)

$R_i < r < R_o$

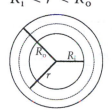

$$\int \vec{E} \cdot d\vec{A} = \frac{q_{encl}}{\varepsilon_0}$$

$$EA = \frac{q_{encl}}{\varepsilon_0}$$

$$E = \frac{q_{encl}}{A\varepsilon_0} = \frac{4\pi R_i^2 \sigma_i}{(4\pi r^2)\varepsilon_0} = \frac{R_i^2 \sigma_i}{r^2 \varepsilon_0}$$

The electric field has a magnitude of $\boxed{\dfrac{R_i^2 \sigma_i}{r^2 \varepsilon_0} \text{ and points radially outward}}$.

Part c)

$r > R_o$

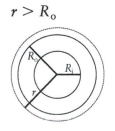

$$\int \vec{E} \cdot d\vec{A} = \frac{q_{encl}}{\varepsilon_0}$$

$$EA = \frac{q_{encl}}{\varepsilon_0}$$

$$E = \frac{q_{encl}}{A\varepsilon_0} = \frac{4\pi R_i^2 \sigma_i - 4\pi R_o^2 \sigma_o}{(4\pi r^2)\varepsilon_0} = \frac{R_i^2 \sigma_i - R_o^2 \sigma_o}{r^2 \varepsilon_0}$$

The magnitude of the electric field is

$$\boxed{\frac{R_i^2\sigma_i - R_o^2\sigma_o}{r^2\varepsilon_0} \text{ and points radially outward if } R_i^2\sigma_i > R_o^2\sigma_o \text{ or inward if } R_i^2\sigma_i < R_o^2\sigma_o.}$$

Part d)

For $r < R_i$, $\boxed{E = 0}$.

For $R_i < r < R_o$, $E = \dfrac{R_i^2\sigma_i}{r^2\varepsilon_0} = \dfrac{(5 \text{ cm})^2\left(20 \times 10^{-6}\, \dfrac{\text{C}}{\text{cm}^2}\right)}{r^2\left(8.85 \times 10^{-12}\, \dfrac{\text{C}^2}{\text{N}\cdot\text{m}^2}\right)} = \boxed{\dfrac{5.65 \times 10^7}{r^2}\, \dfrac{\text{N}\cdot\text{m}^2}{\text{C}}}$

For $r > R_o$, $E = \left|\dfrac{R_i^2\sigma_i - R_o^2\sigma_o}{r^2\varepsilon_0}\right|$

$$= \left|\frac{(5 \text{ cm})^2\left(20 \times 10^{-6}\, \dfrac{\text{C}}{\text{cm}^2}\right) - (8 \text{ cm})^2\left(14 \times 10^{-6}\, \dfrac{\text{C}}{\text{cm}^2}\right)}{r^2\left(8.85 \times 10^{-12}\, \dfrac{\text{C}^2}{\text{N}\cdot\text{m}^2}\right)}\right| = \boxed{\frac{4.47 \times 10^7}{r^2}\, \frac{\text{N}\cdot\text{m}^2}{\text{C}}}$$

REFLECT

The negative value inside the absolute value symbols in part (d) means the electric field points radially inward (*i.e.*, in the negative $\hat{r}$ direction).

Chapter 17
Electrostatics II

Conceptual Questions

17.5 This statement makes sense only if the zero point of the electric potential has been previously defined.

Get Help: Picture It – Electric Potential

17.9 Part a) Yes, a region of constant potential must have zero electric field.

Part b) No, if the electric field is zero, the potential need only be constant.

17.15 The charge on the capacitor remains constant, while the capacitance decreases. Therefore, the energy stored in the capacitor increases.

Get Help: P'Cast 17.5 – Defibrillator

Multiple-Choice Questions

17.21 **D** (its potential energy decreases and its electric potential decreases). The electric field points in the direction of the force a positive charge would experience. A positive charge moving in this direction would lower its potential energy. The electric field also points towards regions of lower potential.

17.27 **B** (doubled).

$$\frac{U_2}{U_1} = \frac{\left(\dfrac{1}{2}\dfrac{Q_2^2}{C_2}\right)}{\left(\dfrac{1}{2}\dfrac{Q_1^2}{C_1}\right)} = \frac{\left(\dfrac{Q^2}{C_2}\right)}{\left(\dfrac{Q^2}{C_1}\right)} = \frac{C_1}{C_2} = \frac{\left(\dfrac{\epsilon_0 A}{d_1}\right)}{\left(\dfrac{\epsilon_0 A}{d_2}\right)} = \frac{d_2}{d_1} = \frac{2d_1}{d_1} = \boxed{2}$$

Get Help: Interactive Example – Parallel Plate Capacitor
P'Cast 17.5 – Defibrillator

Estimation/Numerical Questions

17.33 The electric potential of an electron located a distance of approximately 5×10^{-11} m from the nucleus of an atom is around 25 V.

Problems

17.37

SET UP

A charge ($q_0 = +2.0$ C) is moved through a potential difference of $V = +9.0$ V. The work done by the electric field is equal to $W_{\text{by } \vec{E}} = -Vq_0$. Assuming the charge starts and ends at rest, the work required to move the charge is $W_{\text{required}} = -W_{\text{by } \vec{E}}$.

SOLVE

$$V = \frac{-W_{\text{by }\vec{E}}}{q_0} = \frac{W_{\text{required}}}{q_0}$$

$$W_{\text{required}} = Vq_0 = (9.0 \text{ V})(2.0 \text{ C}) = \boxed{18 \text{ J}}$$

REFLECT

It should require energy to move a positive charge to an area of higher potential.

17.45

SET UP

We are asked to draw the equipotential lines and electric field lines for two pairs of charges—two $+q$ and two $-q$. The equipotential lines will be the closest together near the charges and spread out as we move away. Electric field lines start on positive charges, end on negative charges, and cannot cross one another.

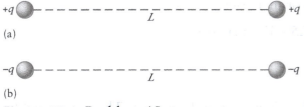

(a)

(b)

Figure 17-1 Problem 45

SOLVE

Parts a) (Equipotential lines) and b) (Electric fields)

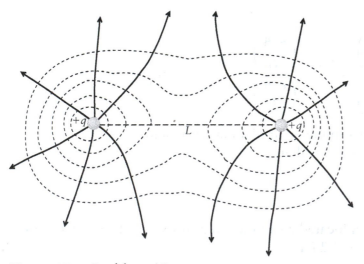

Figure 17-2 Problem 45

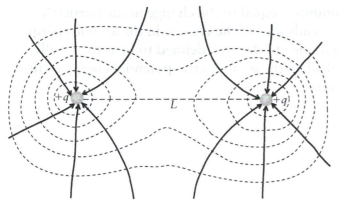

Figure 17-3 Problem 45

REFLECT
Although the shape of the equipotential curves are the same in both setups, the potentials those lines correspond to would not be the same.

17.51

SET UP
Two point charges are placed on the x-axis: $q_1 = 0.5 \ \mu C$ is located at $x = 0$ and $q_2 = -0.2 \ \mu C$ is located at $x = 10$ cm. The electric potential due to a point charge q as a function of distance is $V(r) = \dfrac{kq}{r}$. We can calculate the position where the electric potential is equal to zero by adding the expressions for the potential due to each charge, setting the sum to zero, and solving for x.

SOLVE

$$V(x) = 0 = \frac{kq_1}{x} + \frac{kq_2}{(x - 10)} \ (x \text{ in cm})$$

$$-\frac{q_1}{x} = \frac{q_2}{(x - 10)}$$

$$x = \frac{10q_1}{q_1 + q_2} = \frac{10(0.5 \ \mu C)}{(0.5 \ \mu C) + (-0.2 \ \mu C)} = \boxed{16.7 \text{ cm}}$$

REFLECT
It makes sense that the position where $V = 0$ is to the right of both charges since the magnitude of q_1 is larger than the magnitude of q_2, q_1 is positive, and q_2 is negative.

17.57

SET UP
Three particles, each with charge q, are located at different corners of a rhombus with sides of length a, a short diagonal of length a, and a long diagonal of length b. The relationship between the sides of the rhombus and its diagonals is: $4a^2 = a^2 + b^2$. The total electric potential energy of the charge distribution is equal to the sum of the pairwise potential energies of the charges, $U_{q_A q_B} = \dfrac{kq_A q_B}{r}$. The required external work to bring a fourth charge

q from infinity to the fourth corner of the rhombus is equal to the change in the particle's mechanical energy. Since the particle starts and ends at rest, the change in the kinetic energy is equal to zero. The potential energy of the particle at infinity is defined to be zero. Finally, we can use our answers to parts a and b to calculate the total electric potential energy of the charge distribution consisting of the four charges.

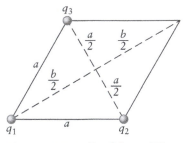

Figure 17-4 Problem 57

SOLVE

Part a)

$$U_{\text{total}} = U_{q_1 q_2} + U_{q_2 q_3} + U_{q_1 q_3} = \frac{kq_1 q_2}{a} + \frac{kq_2 q_3}{a} + \frac{kq_1 q_3}{a} = \frac{k}{a}[q^2 + q^2 + q^2] = \boxed{\frac{3kq^2}{a}}$$

Part b)

$$W = \Delta U = U_f - U_i = U_f - 0 = U_{q_1 q_4} + U_{q_2 q_4} + U_{q_3 q_4} = \frac{kq_1 q_4}{b} + \frac{kq_2 q_4}{a} + \frac{kq_3 q_4}{a}$$

$$= \frac{kq^2}{b} + \frac{kq^2}{a} + \frac{kq^2}{a} = \frac{kq^2}{\sqrt{4a^2 - a^2}} + \frac{2kq^2}{a} = \boxed{\left(2 + \frac{1}{\sqrt{3}}\right)\frac{kq^2}{a}}$$

Part c)

$$U_{\text{total}} = (U_{q_1 q_2} + U_{q_2 q_3} + U_{q_1 q_3}) + (U_{q_1 q_4} + U_{q_2 q_4} + U_{q_3 q_4}) = \left(\frac{3kq^2}{a}\right) + \left(\left(2 + \frac{1}{\sqrt{3}}\right)\frac{kq^2}{a}\right)$$

$$= \boxed{\left(5 + \frac{1}{\sqrt{3}}\right)\frac{kq^2}{a}}$$

REFLECT

We can talk about the electric potential due to one charge, but the electric potential energy due to the interaction between two charges.

17.63

SET UP

A parallel plate capacitor has square plates ($L = 1.0$ m). We can use the expression for the capacitance of a parallel plate capacitor, $C = \frac{\varepsilon_0 A}{d}$, to calculate the separation distance d required to give $C = 8850 \times 10^{-12}$ F.

SOLVE

$$C = \frac{\varepsilon_0 A}{d} = \frac{\varepsilon_0 L^2}{d}$$

$$d = \frac{\varepsilon_0 L^2}{C} = \frac{\left(8.85 \times 10^{-12}\frac{F}{m}\right)(1.0 \text{ m})^2}{8850 \times 10^{-12} \text{ F}} = 0.0010 \text{ m} = \boxed{1.0 \text{ mm}}$$

REFLECT

The dielectric constant of air is $\kappa = 1.00058$, which is approximately 1.

Get Help: Picture It – Capacitance

P'Cast 17.3 – Parallel Plates

17.69

SET UP

The capacitor ($C = 20 \times 10^{-6}$ F) in a defibrillator has a voltage of 10×10^3 V. The energy released into the patient is equal to the potential energy initially stored by the capacitor, $U = \frac{1}{2}CV^2$.

SOLVE

$$U = \frac{1}{2}CV^2 = \frac{1}{2}(20 \times 10^{-6} \text{ F})(10 \times 10^3 \text{ V})^2 = \boxed{1000 \text{ J}}$$

REFLECT

We assumed that all of the energy stored in the capacitor is released into the patient.

Get Help: Interactive Example – Parallel Plate Capacitor

P'Cast 17.5 – Defibrillator

17.73

SET UP

Two capacitors, C_1 and C_2, are arranged in series and in parallel. The equivalent series capacitance is $C_{\text{series}} = 2.00 \ \mu$F, and the equivalent parallel capacitance is $C_{\text{parallel}} = 8.00 \ \mu$F. We can solve for the values of C_1 and C_2 using the expressions for the equivalent series $\left(\frac{1}{C_{\text{series}}} = \frac{1}{C_1} + \frac{1}{C_2}\right)$ and parallel ($C_{\text{parallel}} = C_1 + C_2$) capacitances.

SOLVE

Equivalent capacitances:

$$\frac{1}{C_{\text{series}}} = \frac{1}{C_1} + \frac{1}{C_2}$$

$$C_{\text{series}} = \frac{C_1 C_2}{C_1 + C_2}$$

$$C_{\text{parallel}} = C_1 + C_2$$

Solving for C_1 and C_2:

$$C_{series}(C_1 + C_2) = C_1 C_2$$

$$C_{series} C_{parallel} = (C_{parallel} - C_2) C_2$$

$$C_2^2 - C_{parallel} C_2 + C_{series} C_{parallel} = 0$$

$$C_2 = \frac{-(-C_{parallel}) \pm \sqrt{(-C_{parallel})^2 - 4(1)(C_{series} C_{parallel})}}{2(1)} = \frac{C_{parallel} \pm \sqrt{C_{parallel}^2 - 4 C_{series} C_{parallel}}}{2}$$

$$= \frac{(8.00\ \mu F) \pm \sqrt{(8.00\ \mu F)^2 - 4(2.00\ \mu F)(8.00\ \mu F)}}{2} = \frac{(8.00\ \mu F) \pm 0}{2} = \boxed{4.00\ \mu F}$$

$$C_1 = C_{parallel} - C_2 = (8.00\ \mu F) - (4.00\ \mu F) = \boxed{4.00\ \mu F}$$

REFLECT

Quickly double-checking our answer, we find that $4 + 4 = 8$, and $(1/4) + (1/4) = (1/2)$, which matches the values listed in the problem statement.

Get Help: Interactive Example – Capacitor Network
P'Cast 17.7 – Multiple Capacitors

17.79

SET UP

Three capacitors—$C_1 = 10.0\ \mu F$, $C_2 = 40.0\ \mu F$, and $C_3 = 100.0\ \mu F$—are connected in series across a 12.0 V battery. The equivalent capacitance of three capacitors in series is $\frac{1}{C_{series}} = \frac{1}{C_1} + \frac{1}{C_2} + \frac{1}{C_3}$. When they are wired up to the battery, the capacitors will store charge. The capacitors are wired in series, which means the charge on each capacitor will be the same. Because of this, we can use the equivalent capacitance of the capacitor network and the potential difference across the battery to calculate the charge on each capacitor. The potential difference $V = \frac{Q}{C}$ across each capacitor will not be equal, though, since the capacitances are different.

SOLVE

Part a)

$$\frac{1}{C_{series}} = \frac{1}{C_1} + \frac{1}{C_2} + \frac{1}{C_3} = \frac{1}{10.0\ \mu F} + \frac{1}{40.0\ \mu F} + \frac{1}{100.0\ \mu F} = \frac{27}{200.0\ \mu F}$$

$$C_{series} = \frac{200.0\ \mu F}{27} = \boxed{7.41\ \mu F}$$

Part b)

$$Q = C_{series} V = (7.41 \times 10^{-6}\ F)(12.0\ V) = 8.89 \times 10^{-5}\ C = \boxed{88.9\ \mu C}$$

Part c)

$$V_1 = \frac{Q}{C_1} = \frac{88.9 \times 10^{-6} \text{ C}}{10.0 \times 10^{-6} \text{ F}} = \boxed{8.89 \text{ V}}$$

$$V_2 = \frac{Q}{C_2} = \frac{88.9 \times 10^{-6} \text{ C}}{40.0 \times 10^{-6} \text{ F}} = \boxed{2.22 \text{ V}}$$

$$V_3 = \frac{Q}{C_3} = \frac{88.9 \times 10^{-6} \text{ C}}{100.0 \times 10^{-6} \text{ F}} = \boxed{0.889 \text{ V}}$$

REFLECT

The sum of the potential differences across each capacitor should be 12.0 V: (8.89 V) + (2.22 V) + (0.889 V) = 12.0 V. The potential energy stored in the three capacitors in series is equal to the potential energy stored by the equivalent capacitor.

Get Help: Interactive Example – Capacitor Network
P'Cast 17.7 – Multiple Capacitors

17.85

SET UP

An air-gap capacitor ($C_i = 2800 \times 10^{-12}$ F) accumulates a charge Q_i when it is connected to a 16 V battery. While the battery is still connected, a dielectric material ($\kappa = 5.8$) is placed between the plates of the capacitor, which increases the capacitance to $C_f = \kappa C_i$. The final charge on the capacitor is Q_f. We can use the definition of capacitance to calculate the amount of charge that flowed from the battery to the capacitor upon adding the dielectric.

SOLVE

$$\Delta Q = Q_f - Q_i = C_f V - C_i V = (\kappa C_i)V - C_i V = C_i V(\kappa - 1)$$

$$= (2800 \times 10^{-12} \text{ F})(16 \text{ V})(5.8 - 1) = \boxed{2.2 \times 10^{-7} \text{ C}}$$

REFLECT

Inserting a dielectric into a capacitor increases its capacitance, which means it can store more charge for a given potential difference. Therefore, charge will flow from the battery onto the plates of the capacitor. If the battery were not connected when the dielectric was inserted, the charge on the plates would remain constant and the potential difference across the capacitor would decrease.

Get Help: Interactive Example – Parallel Plate Capacitor
P'Cast 17.9 – Capacitance of Membranes

17.89

SET UP

A uniform electric field of $E = 2.00 \times 10^3$ N/C points towards $+x$. The potential difference $V_b - V_a$ between $x_a = -0.300$ m and $x_b = 0.500$ m is given by $V_b - V_a = -E(x_b - x_a)$. A positive test charge $q_0 = +2.00 \times 10^{-9}$ C is released from rest at point a. We can use conservation of energy to calculate the kinetic energy of the charge when it passes through point b; the change in the charge's potential energy is equal to $q_0 \Delta V$. If the charge were negative rather than positive, the potential difference between points a and b would be the same since this only depends upon the location of a and b in space, but the negative charge would be accelerated towards $-x$ by the electric field and not pass through point b.

SOLVE

Part a)

$$V_b - V_a = -E(x_b - x_a) = -\left(2.00 \times 10^3 \frac{\text{N}}{\text{C}}\right)((0.500 \text{ m}) - (-0.300 \text{ m})) = \boxed{-1.60 \times 10^3 \text{ V}}$$

Part b)

$$\Delta U + \Delta K = q_0 \Delta V + (K_b - K_a) = q_0 \Delta V + (K_b - 0) = 0$$

$$K_b = -q_0 \Delta V = -(2.00 \times 10^{-9} \text{ C})(-1.60 \times 10^3 \text{ V}) = \boxed{3.2 \times 10^{-6} \text{ J}}$$

Part c)
The potential difference between the two points would remain the same. However, if a negative charge were placed at rest at point a then it would never reach point b unless an external force acted upon it. The charge would accelerate in the $-x$ direction away from point b.

REFLECT
Use your intuition to help keep your signs straight. The electric field in this problem points towards $+x$, which means point b must be at a lower potential than point a. Therefore, $V_b - V_a$ must be a negative number.

17.95

SET UP

An infinitely long line charge $\left(\lambda = 2.00 \times 10^{-9} \frac{\text{C}}{\text{m}}\right)$ lies along the z-axis. We can use Gauss' law to first find the electric field due to this line charge, integrate the field with respect to r to get an expression for the potential as a function of r, and then calculate the potential difference between point b at (40.0 cm, 30.0 cm, 0) and point c at (200 cm, 0, 50.0 cm). Since we're finding the potential difference between two specific points, the reference point where $V = 0$ is irrelevant. The equipotential surfaces are shaped like circular cylinders coaxial with the line charge.

SOLVE

Part a)
Gauss' law:

$$\oint \vec{E} \cdot d\vec{A} = \frac{q_{\text{encl}}}{\varepsilon_0}$$

$$E(2\pi r L) = \frac{\lambda L}{\varepsilon_0}$$

$$E = \frac{\lambda}{2\pi \varepsilon_0 r}$$

Potential difference:

$$V(r) = -\int \vec{E} \cdot d\vec{s} = -\int \left(\frac{\lambda}{2\pi\varepsilon_0 r}\right) dr = -\frac{\lambda}{2\pi\varepsilon_0}\ln(r) + C$$

$$V_c - V_b = -\frac{\lambda}{2\pi\varepsilon_0}\ln(r_c) + \frac{\lambda}{2\pi\varepsilon_0}\ln(r_b) = \frac{\lambda}{2\pi\varepsilon_0}\ln\left(\frac{r_b}{r_c}\right) = \frac{\lambda}{2\pi\varepsilon_0}\ln\left(\frac{\sqrt{x_b^2 + y_b^2 + z_b^2}}{\sqrt{x_c^2 + y_c^2 + z_c^2}}\right)$$

$$= \frac{\left(2.00 \times 10^{-9}\frac{C}{m}\right)}{2\pi\left(8.85 \times 10^{-12}\frac{C^2}{N \cdot m^2}\right)}\ln\left(\frac{\sqrt{(40.0 \text{ cm})^2 + (30.0 \text{ cm})^2 + (0)^2}}{\sqrt{(200 \text{ cm})^2 + (0)^2 + (50.0 \text{ cm})^2}}\right)$$

$$= \frac{\left(2.00 \times 10^{-9}\frac{C}{m}\right)}{2\pi\left(8.85 \times 10^{-12}\frac{C^2}{N \cdot m^2}\right)}\ln\left(\frac{50.0 \text{ cm}}{206.2 \text{ cm}}\right) = \boxed{-51.0 \text{ V}}$$

Part b) Our answer is independent of the location of $V = 0$ since we're interested in the potential difference between two points.

Part c) The equipotential surfaces are the surfaces of cylinders coaxial with the line charge.

REFLECT
We can't use $r = 0$ or $r = $ infinity as the reference point where $V = 0$ for an infinite line of charge because the natural logarithm diverges at both 0 and infinity. Therefore, we usually pick and arbitrary distance R_{ref} to define $V = 0$.

17.103

SET UP
We are asked to find the equivalent capacitance of four capacitors—$C_1 = 3$ pF, $C_2 = 6$ pF, $C_3 = 4$ pF, $C_4 = 6$ pF—wired up as shown (see figure). The potential difference over the entire top branch (made up of C_3 and C_4) is equal to the potential difference over the entire bottom branch (made up of C_1 and C_2), which means these branches are in parallel with one another. The capacitors in each branch (C_1 and C_2, C_3 and C_4) are wired in series since the current through each branch is the same throughout the branch. We can find the overall equivalent capacitance by first finding the equivalent capacitance of each branch and then the equivalent capacitance of the two branches together. The equivalent capacitance of two capacitors in series is $\frac{1}{C_{\text{series}}} = \frac{1}{C_1} + \frac{1}{C_2}$; the equivalent capacitance of two capacitors in parallel is $C_{\text{parallel}} = C_1 + C_2$.

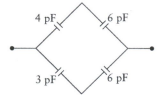

Figure 17-5 Problem 103

SOLVE
Redrawn diagram:

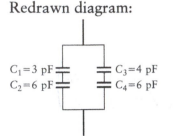

Figure 17-6 Problem 103

Equivalent capacitance:

$$\frac{1}{C_{12}} = \frac{1}{C_1} + \frac{1}{C_2} = \frac{1}{3 \text{ pF}} + \frac{1}{6 \text{ pF}} = \frac{3}{6 \text{ pF}}$$

$$C_{12} = 2 \text{ pF}$$

$$\frac{1}{C_{34}} = \frac{1}{C_3} + \frac{1}{C_4} = \frac{1}{4 \text{ pF}} + \frac{1}{6 \text{ pF}} = \frac{5}{12 \text{ pF}}$$

$$C_{34} = \frac{12}{5} \text{pF}$$

$$C_{1234} = C_{12} + C_{34} = (2 \text{ pF}) + \left(\frac{12}{5}\text{pF}\right) = \boxed{\frac{22}{5}\text{pF} = 4.4 \text{ pF}}$$

REFLECT
Redrawing the circuit diagram using right angles helps when determining which elements are in series and which are in parallel.

Get Help: Interactive Example – Capacitor Network
P'Cast 17.7 – Multiple Capacitors

17.109

SET UP
A spherical, air-filled capacitor consists of an inner conductor of radius R_1 and an outer conductor of radius R_2. We can use Gauss' law to find the expression for the electric field in the region $R_1 < r < R_2$ and then integrate it to find the potential difference V between the conductors. Finally, we can rearrange the expression for Q and compare it to $Q = CV$ in order to determine the capacitance. Once we have an expression for C, we can determine the effect that increasing R_1 has on the capacitance.

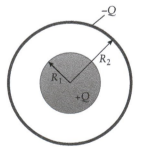

Figure 17-7 Problem 109

SOLVE

Part a)

Gauss' law:

$$\oint \vec{E} \cdot d\vec{A} = \frac{q_{encl}}{\varepsilon_0}$$

$$E(4\pi r^2) = \frac{+Q}{\varepsilon_0}$$

$$E = \frac{Q}{4\pi\varepsilon_0 r^2}$$

Potential difference:

$$V = -\int_{R_2}^{R_1} \vec{E} \cdot d\vec{s} = -\int_{R_2}^{R_1} E \, dr = -\int_{R_2}^{R_1} \left(\frac{Q}{4\pi\varepsilon_0 r^2}\right) dr = \frac{Q}{4\pi\varepsilon_0}\left[\frac{1}{r}\right]_{R_2}^{R_1}$$

$$= \frac{Q}{4\pi\varepsilon_0}\left[\frac{1}{R_1} - \frac{1}{R_2}\right] = \frac{Q}{4\pi\varepsilon_0}\left[\frac{R_2 - R_1}{R_1 R_2}\right]$$

Capacitance:

$$V = \frac{Q}{4\pi\varepsilon_0}\left[\frac{R_2 - R_1}{R_1 R_2}\right]$$

$$Q = 4\pi\varepsilon_0\left[\frac{R_1 R_2}{R_2 - R_1}\right]V = CV$$

$$\boxed{C = 4\pi\varepsilon_0\left[\frac{R_1 R_2}{R_2 - R_1}\right]}$$

Part b) Since the quantity $R_2 - R_1$ in the denominator will get small, the capacitance will increase, which means the capacitor will store more charge per volt. In addition, the numerator will also get larger, which will also increase the capacitance.

REFLECT

It doesn't matter if the inner conductor is solid or hollow. The charge enclosed by a Gaussian surface with radius $R_1 < r < R_2$ is the same in either case. Also, the expression for the capacitance of a spherical capacitor mimics the capacitance of a parallel plate capacitor—the numerator consists of an area multiplied by ε_0 and the denominator is a measure of the distance between the plates.

17.115

SET UP

A parallel plate capacitor has an area A and separation d. A conducting slab of thickness d' is inserted between, and parallel to, the plates. This set up is equivalent to two parallel plate capacitors in series. If we define the distances of the gaps between the plates as s_1 and s_2, the separation distances for each "sub-capacitor" are $(d - s_1 - d')$ and $(d - s_2 - d')$, where $d = s_1 + s_2 + d'$. We can use the general expression for the capacitance of a parallel plate capacitor, $C = \dfrac{\varepsilon_0 A}{d}$, and the equivalent capacitance of two capacitors in series, $\dfrac{1}{C_{equiv}} = \dfrac{1}{C_1} + \dfrac{1}{C_2}$, to calculate the capacitance of the object upon adding the conducting slab.

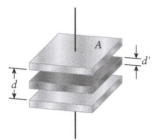

Figure 17-8 Problem 115

SOLVE

Part a)

Original capacitance:

$$C_{before} = \frac{\varepsilon_0 A}{d}$$

Adding the conducting slab:

$$\frac{1}{C_{after}} = \frac{1}{C_1} + \frac{1}{C_2} = \frac{1}{\left(\dfrac{\varepsilon_0 A}{d - s_1 - d'}\right)} + \frac{1}{\left(\dfrac{\varepsilon_0 A}{d - s_2 - d'}\right)} = \frac{d - s_1 - d' + d - s_2 - d'}{\varepsilon_0 A}$$

$$= \frac{2d - d' - (s_1 + s_2 + d')}{\varepsilon_0 A} = \frac{2d - d' - (d)}{\varepsilon_0 A} = \frac{d - d'}{\varepsilon_0 A}$$

$$\boxed{C_{after} = \frac{\varepsilon_0 A}{d - d'}}$$

The capacitance has increased upon adding the conducting slab.

Part b) Since our expression is independent of s_1 and s_2, the effect is independent of the location of the slab.

REFLECT

Even though we treated the "sub-capacitors" as being in series, the capacitance has *increased* upon adding the conducting slab. This may seem counterintuitive—adding identical capacitors in series lowers their equivalent capacitance—but, keep in mind, the "sub-capacitors" are *not identical to the original capacitor*! These "sub-capacitors" have the same cross-sectional area, but a *smaller* separation distance than the original capacitor.

Get Help: Interactive Example – Capacitor Network
P'Cast 17.7 – Multiple Capacitors

Chapter 18
Moving Charges

Conceptual Questions

18.3   There is no contradiction. If a conductor is in electrostatic equilibrium, the electric field with it must be zero. However, any conductor that is carrying a current is definitely not in electrostatic equilibrium.

18.7   A small resistance will dissipate more power and generate more heat.

18.11 The voltage drop across each bulb in the old series string was about 1/50 of 110 V, or 2.2 V. The modern parallel connection puts the full 110 V across each bulb. Placing 110 V across one of the old bulbs, designed to operate at 2.2 V, would result in excessive current in the filament, which would burn out the bulb immediately, perhaps in a spectacular manner.

Get Help: P'Cast 18.7 – Resistors in Combination

Multiple-Choice Questions

18.17 A (increases along length of the wire). The drift speed is inversely proportional to the cross-sectional area of the wire.

Get Help: P'Cast 18.2 – How Fast?

18.21 D (2 A). Since we are using the same wire, the resistance remains constant:

$$\frac{V_2}{V_1} = \frac{i_2 R}{i_1 R}$$

$$i_2 = \left(\frac{V_2}{V_1}\right) i_1 = \left(\frac{2\text{ V}}{1\text{ V}}\right)(1\text{ A}) = 2\text{ A}$$

18.25 B (decreases). A light bulb acts as a resistor and the equivalent resistance for two resistors in parallel is $\dfrac{1}{R_{\text{equiv}}} = \dfrac{1}{R_1} + \dfrac{1}{R_2}$.

Get Help: P'Cast 18.7 – Resistors in Combination

Estimation/Numerical Questions

18.33 The resistor is about 1 kΩ and the capacitor is about 100 μF. This gives a time constant of 0.1 s, which is on the smaller side.

Get Help: Interactive Example – RC III

Problems

18.37

SET UP

A synchrotron facility creates an electron beam with a current $i = 0.487$ A. We can calculate the number of electrons that pass a given point in an hour from the current and the magnitude of the charge on an electron, 1.6×10^{-19} C.

SOLVE

$$1 \text{ hr} \times \frac{3600 \text{ s}}{1 \text{ hr}} \times \frac{0.487 \text{ C}}{1 \text{ s}} \times \frac{1 \text{ electron}}{1.6 \times 10^{-19} \text{ C}} = \boxed{1.09 \times 10^{22} \text{ electrons}}$$

REFLECT

We were given the current, so we didn't need to use the dimensions of the synchrotron or the speed of the electrons.

18.45

SET UP

A length of wire has an initial length $L_1 = 8$ m and resistance $R_1 = 4\ \Omega$. The wire is then stretched uniformly to a final length $L_2 = 16$ m. Since the mass and density of the wire must be constant, the volume of the wire must also remain constant. Therefore, the cross-sectional area will decrease upon stretching. We can then calculate the new resistance of the wire R_2 using the definition of R in terms of the resistivity, $R = \dfrac{\rho L}{A}$.

SOLVE

New cross-sectional area:

$$A_1 L_1 = A_2 L_2$$

$$A_1(8 \text{ m}) = A_2(16 \text{ m})$$

$$A_2 = \frac{A_1}{2}$$

New resistance:

$$\frac{R_2}{R_1} = \frac{\left(\dfrac{\rho L_2}{A_2}\right)}{\left(\dfrac{\rho L_1}{A_1}\right)} = \frac{L_2 A_1}{A_2 L_1} = \frac{L_2 A_1}{\left(\dfrac{A_1}{2}\right) L_1} = \frac{2L_2}{L_1} = \frac{2(16 \text{ m})}{8 \text{ m}} = 4$$

$$R_2 = 4R_1 = 4(4\ \Omega) = \boxed{16\ \Omega}$$

REFLECT

Looking at the expression for the resistance in terms of the resistivity, $R = \dfrac{\rho L}{A}$, increasing the length of the wire and decreasing the cross-sectional area both cause the resistance to increase.

Get Help: P'Cast 18.3 – Stretched Wire

18.51

SET UP

A thin, conducting, spherical shell has an inner radius r_i and an outer radius r_o. The conductor has a resistivity ρ. If we split the shell into thin concentric spheres of thickness dr, the total resistance of the shell is the sum of the resistances due to each one, $\dfrac{\rho dr}{A}$, where the surface area of a sphere $A = 4\pi r^2$. We can accomplish this sum through integration, $R = \displaystyle\int_{r_i}^{r_o} \dfrac{\rho dr}{A}$.

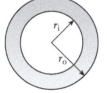

Figure 18-1 Problem 51

SOLVE

$$R = \int_{r_i}^{r_o} \frac{\rho dr}{A} = \rho \int_{r_i}^{r_o} \frac{dr}{4\pi r^2} = \frac{\rho}{4\pi}\left[-\frac{1}{r} \right]_{r_i}^{r_o} = \frac{\rho}{4\pi}\left[-\frac{1}{r_o} + \frac{1}{r_i} \right] = \boxed{\frac{\rho}{4\pi}\left[\frac{1}{r_i} - \frac{1}{r_o} \right]}$$

REFLECT

Resistivity is associated with a material (*e.g.*, copper) while resistance is associated with an object (*e.g.*, a 1-m-long copper wire).

18.57

SET UP

A light bulb draws a current of $i = 1.0$ A when connected to a voltage of $V = 12$ V. The resistance of the filament is given by $V = iR$.

SOLVE

$$V = iR$$

$$R = \frac{V}{i} = \frac{12 \text{ V}}{1.0 \text{ A}} = \boxed{12 \ \Omega}$$

REFLECT

A material with a constant resistance over a range of applied potential is said to be ohmic.

18.63

SET UP

Two resistors, $R_9 = 9.0 \ \Omega$ and $R_3 = 3.0 \ \Omega$, are wired in series across a battery ($V = 9.0$ V). Since the resistors are connected together in series, the current through each of them is the same. We can calculate the current from the battery voltage and the equivalent resistance of the two resistors in series. The equivalent resistance for the resistors in series is $R_{\text{equiv}} = R_9 + R_3$. Even though the current is the same through both resistors, the voltage drop across each one will be different. We can calculate the voltage drops V_9 and V_3 from the current found in part a and the individual resistances.

SOLVE

Part a)

Equivalent resistance:

$$R_{equiv} = R_9 + R_3 = (9.0 \ \Omega) + (3.0 \ \Omega) = 12.0 \ \Omega$$

Current:

$$V = iR_{equiv}$$

$$i = \frac{V}{R_{equiv}} = \frac{9.0 \ V}{12.0 \ \Omega} = \boxed{0.75 \ A}$$

Part b)

$$V_9 = iR_9 = (0.75 \ A)(9.0 \ \Omega) = \boxed{6.75 \ V}$$

$$V_3 = iR_3 = (0.75 \ A)(3.0 \ \Omega) = \boxed{2.25 \ V}$$

REFLECT

The sum of the voltages across the two resistors must equal the voltage of the battery:

$$V_9 + V_3 \overset{?}{=} V$$

$$(6.75 \ V) + (2.25 \ V) = 9.0 \ V$$

18.69

SET UP

Two resistors A and B (resistances R_A and R_B) are connected in series to a battery ($V = 6.0$ V) and the voltage across resistor A is $V_{A, s} = 4.0$ V. The circuit is disconnected and rewired such that A and B are in parallel with the battery. The current through resistor B in this setup is $i_{B, p} = 2.0$ A. We can determine R_B from the information related to the parallel circuit. In this setup, the voltage across each resistor must be equal to the voltage across the battery. Since we know the voltage across and the current passing through resistor B we can calculate R_B. Now that we know the value for R_B, we can calculate R_A from the information related to the series circuit. First, the sum of the voltages across A and B must equal the voltage across the battery; this will provide us with the voltage across B in the series circuit. We already know R_B, so we can calculate the current in that circuit, which is the same through each element because they are all in series. Finally, by dividing the voltage across A by the series current, we will get R_A.

SOLVE

Resistors in parallel:

$$V_{B, p} = V_{A, p} = V$$

$$V_{B, p} = i_{B, p} R_B$$

$$R_B = \frac{V_{B, p}}{i_{B, p}} = \frac{V}{i_{B, p}} = \frac{6.0 \text{ V}}{2.0 \text{ A}} = \boxed{3.0 \text{ } \Omega}$$

Resistors in series:

$$V = V_{A, s} + V_{B, s}$$

$$V_{B, s} = V - V_{A, s} = (6.0 \text{ V}) - (4.0 \text{ V}) = 2.0 \text{ V}$$

$$V_{B, s} = i_s R_B$$

$$i_s = \frac{V_{B, s}}{R_B} = \frac{2.0 \text{ V}}{3.0 \text{ } \Omega} = 0.667 \text{ A}$$

$$V_{A, s} = i_s R_A$$

$$R_A = \frac{V_{A, s}}{i_s} = \frac{4.0 \text{ V}}{0.667 \text{ } \Omega} = \boxed{6.0 \text{ A}}$$

REFLECT

For problems dealing with networks of resistors and capacitors, it's easiest to split the problem up into smaller steps rather than trying tackle it all at once. Also, realizing what remains constant in a given circuit—voltage for elements in parallel, current for elements in series—will usually help start you off.

18.77

SET UP

The total energy required to run a 1500 W heater for 8 hours is equal to the product of the power and the time interval. Since we are given the cost the power company charges you in kWh ($0.11/kWh), we can multiply the numbers directly and then convert into kWh to find the total cost.

SOLVE

$$(1500 \text{ W})(8 \text{ h}) \times \frac{1 \text{ kW} \cdot \text{h}}{1000 \text{ W} \cdot \text{h}} \times \frac{\$0.11}{1 \text{ kW} \cdot \text{h}} = \boxed{\$1.32}$$

REFLECT

A kilowatt-hour is a unit of energy commonly used in electricity bills.

Get Help: P'Cast 18.11 – Drying Your Hair

18.79

SET UP

A resistor $R = 4.00 \times 10^6 \ \Omega$ and a capacitor $C = 3.00 \times 10^{-6} \ \text{F}$ are wired in series with a power supply. The time constant τ for this circuit is equal to the product RC.

SOLVE

$$\tau = RC = (4.00 \times 10^6 \ \Omega)(3.00 \times 10^{-6} \ \text{F}) = \boxed{12.0 \ \text{s}}$$

REFLECT

The prefix *mega-* corresponds to 10^6 and *micro-* corresponds to 10^{-6}, so if we multiply them together the factors of 10 will cancel.

18.83

SET UP

A capacitor ($C = 12.5 \times 10^{-6} \ \text{F}$) is first charged to a potential $V = 50.0 \ \text{V}$ and then discharged through a resistor ($R = 75.0 \ \Omega$). The charge on a discharging capacitor as a function of time is described by $q(t) = q_{max} e^{-\frac{t}{RC}}$. The energy stored in a capacitor is $U = \frac{q^2}{2C}$. We can use these two expressions to calculate the time at which the charge or energy, respectively, reach 10% of their initial value. The current at those times can be calculated using the expression for the current in a discharging series RC circuit as a function of time, $i(t) = -\frac{V}{R} e^{-\frac{t}{RC}}$, where V is the initial voltage across the capacitor.

SOLVE

Part a)

i) Charge:

$$q(t) = q_{max} e^{-\frac{t}{RC}} = 0.100 q_{max}$$

$$t = -RC\ln(0.100) = -(75.0 \ \Omega)(12.5 \times 10^{-6} \ \text{F})\ln(0.100) = \boxed{2.2 \times 10^{-3} \ \text{s}}$$

ii) Energy:

$$U(t) = \frac{1}{2C}(q(t))^2 = \frac{1}{2C}\left(q_{max} e^{-\frac{t}{RC}}\right)^2 = \frac{q_{max}^2}{2C}\left(e^{-\frac{2t}{RC}}\right) = U_{max}\left(e^{-\frac{2t}{RC}}\right)$$

$$U(t) = 0.100 U_{max} = U_{max}\left(e^{-\frac{2t}{RC}}\right)$$

$$t = -\frac{RC}{2}\ln(0.100) = -\frac{(75.0 \ \Omega)(12.5 \times 10^{-6} \ \text{F})}{2}\ln(0.100) = \boxed{1.1 \times 10^{-3} \ \text{s}}$$

Part b)

Current at $t = 2.2 \times 10^{-3}$ s:

$$|i(2.2 \times 10^{-3} \text{ s})| = \left| -\frac{50.0 \text{ V}}{75.0 \text{ }\Omega} e^{-\frac{2.2 \times 10^{-3} \text{ s}}{(75.0 \text{ }\Omega)(12.5 \times 10^{-6} \text{ F})}} \right| = \boxed{0.067 \text{ A}}$$

Current at $t = 1.1 \times 10^{-3}$ s:

$$|i(1.1 \times 10^{-3} \text{ s})| = \left| -\frac{50.0 \text{ V}}{75.0 \text{ }\Omega} e^{-\frac{1.1 \times 10^{-3} \text{ s}}{(75.0 \text{ }\Omega)(12.5 \times 10^{-6} \text{ F})}} \right| = \boxed{0.21 \text{ A}}$$

REFLECT

The time constant $\tau = RC$ allows us to compare two RC circuits and make judgments, for example, on which one charges "faster" or "slower." A discharging capacitor loses about 37% of its initial charge by $t = \tau$.

Get Help: Interactive Example – RC III

18.89

SET UP

The electric charge on a wayward satellite as a function of charge is described by $q(t) = \left(\frac{100 \text{ }\mu\text{C}}{\text{s}^2}\right)t^2 + \left(\frac{150 \text{ }\mu\text{C}}{\text{s}}\right)t + (28 \text{ }\mu\text{C})$. We can evaluate this function at $t = 0$ and $t = 2.5$ s in order to find the amount of charge the satellite possesses at those times. The rate at which charge flows onto the satellite is equal to the derivative of the charge with respect to time. Evaluating this derivative at $t = 0$ and $t = 12$ s will give the rates at those times.

SOLVE

Part a)

$$q(0) = \left(\frac{100 \text{ }\mu\text{C}}{\text{s}^2}\right)(0)^2 + \left(\frac{150 \text{ }\mu\text{C}}{\text{s}}\right)(0) + (28 \text{ }\mu\text{C}) = \boxed{28 \text{ }\mu\text{C}}$$

Part b)

$$q(2.5 \text{ s}) = \left(\frac{100 \text{ }\mu\text{C}}{\text{s}^2}\right)(2.5 \text{ s})^2 + \left(\frac{150 \text{ }\mu\text{C}}{\text{s}}\right)(2.5 \text{ s}) + (28 \text{ }\mu\text{C}) = \boxed{1028 \text{ }\mu\text{C}}$$

Part c)

Rate at which charge flows as a function of time:

$$\frac{dq}{dt} = \frac{d}{dt}\left[\left(\frac{100 \text{ }\mu\text{C}}{\text{s}^2}\right)t^2 + \left(\frac{150 \text{ }\mu\text{C}}{\text{s}}\right)t + (28 \text{ }\mu\text{C})\right] = 2\left(\frac{100 \text{ }\mu\text{C}}{\text{s}^2}\right)t + \left(\frac{150 \text{ }\mu\text{C}}{\text{s}}\right)$$

Rate of charge flow at $t = 0$:

$$\left.\frac{dq}{dt}\right|_{t=0} = 2\left(\frac{100\ \mu C}{s^2}\right)(0) + \left(\frac{150\ \mu C}{s}\right) = \boxed{150\frac{\mu C}{s}}$$

Part d)

$$\left.\frac{dq}{dt}\right|_{t=12\ s} = 2\left(\frac{100\ \mu C}{s^2}\right)(12\ s) + \left(\frac{150\ \mu C}{s}\right) = \boxed{2550\frac{\mu C}{s}}$$

REFLECT

The functional form for $q(t)$ is an upward-facing parabola, which means the amount of charge at $t = 0$ will be a minimum for $t \geq 0$. The slope of the parabola increases linearly with t, so the rate of charge flow at $t = 0$ will also be a minimum.

Get Help: P'Cast 18.1 – Charging Sphere

18.97

SET UP

Three resistors—$R_1 = 20\ \Omega$, $R_2 = 12\ \Omega$, $R_3 = 8\ \Omega$—are wired together: resistors R_1 and R_2 are parallel to one another and this setup is connected in series to R_3. The voltage across the entire resistor network is $V = 5.00$ V. We can calculate the equivalent total resistance by first calculating the equivalent resistance R_{12} for R_1 and R_2 in parallel and then the total equivalent resistance R_{123} for R_{12} and R_3 in series. Since R_3 is in series with R_{12}, the current in R_3 will be equal to the current in R_{123}, which is equal to $\dfrac{V}{R_{123}}$. We can then use the current through R_3 and V to calculate the voltage across R_1 and R_2 in parallel. Once we know the voltage across each resistor R_1 and R_2, we can calculate the current through each one. Finally, the power dissipated by each resistor can be calculated directly from the current and voltage, $P = iV$.

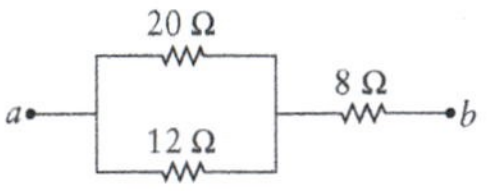

Figure 18-2 Problem 97

SOLVE

Part a)

$$\frac{1}{R_{12}} = \frac{1}{R_1} + \frac{1}{R_2} = \frac{1}{20\ \Omega} + \frac{1}{12\ \Omega} = \frac{8}{60\ \Omega}$$

$$R_{12} = \frac{60}{8}\Omega = 7.5\ \Omega$$

$$R_{123} = R_{12} + R_3 = (7.5\ \Omega) + (8\ \Omega) = \boxed{15.5\ \Omega}$$

Part b)

Current through R_3:

$$V = i_3 R_{123}$$

$$i_3 = \frac{V}{R_{123}} = \frac{5.00 \text{ V}}{15.5 \text{ }\Omega} = \boxed{0.32 \text{ A}}$$

Voltage across R_1 and R_2:

$$V = V_{12} + V_3$$

$$V_{12} = V - V_3 = V - i_3 R_3 = (5.00 \text{ V}) - (0.32 \text{ A})(8 \text{ }\Omega) = 2.44 \text{ V}$$

Current through R_1:

$$V_{12} = i_1 R_1$$

$$i_1 = \frac{V_{12}}{R_1} = \frac{2.44 \text{ V}}{20 \text{ }\Omega} = \boxed{0.12 \text{ A}}$$

Current through R_2:

$$V_{12} = i_2 R_2$$

$$i_2 = \frac{V_{12}}{R_2} = \frac{2.44 \text{ V}}{12 \text{ }\Omega} = \boxed{0.20 \text{ A}}$$

Part c)

Power dissipated by R_1:

$$P_1 = i_1 V_{12} = (0.12 \text{ A})(2.44 \text{ V}) = \boxed{0.29 \text{ W}}$$

Power dissipated by R_2:

$$P_2 = i_2 V_{12} = (0.20 \text{ A})(2.44 \text{ V}) = \boxed{0.49 \text{ W}}$$

Power dissipated by R_3:

$$P_3 = i_3 V_3 = i_3(i_3 R_3) = i_3^2 R_3 = (0.32 \text{ A})^2(8 \text{ }\Omega) = \boxed{0.83 \text{ W}}$$

Part d) The power dissipated by R_1 is less than the power dissipated by the other two resistors because the current through and voltage across R_1 are lower than the other resistors, and $P = iV$.

REFLECT

Remember that two circuit elements in parallel have the same voltage, and two circuit elements in series have the same current.

Get Help: P'Cast 18.9 – A Flashlight

18.103

SET UP

The energy stored in a capacitor is $U = \dfrac{q^2}{2C}$. The charge on a charging capacitor as a function of time in a series RC circuit is described by $q(t) = CV(1 - e^{-\frac{t}{RC}})$. Combining these two expressions will give us an expression describing the time dependence of the energy stored in a charging capacitor in an RC circuit.

SOLVE

$$U(t) = \frac{1}{2C}(q(t))^2 = \frac{1}{2C}(CV(1 - e^{-\frac{t}{RC}}))^2 = \boxed{\frac{CV^2}{2}(1 - e^{-\frac{t}{RC}})^2}$$

REFLECT

At $t = 0$, there is no charge on the capacitor, and the energy stored in it will be 0, as expected. As t approaches infinity, the capacitor should be fully charged and attain its maximum stored energy of $U_{max} = \dfrac{q^2_{max}}{2C} = \dfrac{CV^2}{2}$.

Get Help: Interactive Example – RC III

Chapter 19
Magnetism

Conceptual Questions

19.1 Use both ends of one iron rod to approach the other iron rods. If both ends of the rod you are holding attracts both ends of the other two rods, then the one you are holding is not magnetized iron.

19.5 Part a) The electric field points into the page.

Part b) The beam is deflected into the page.

Part c) The electron beam is not deflected.

19.11 Since the directions of the two currents are opposite, the magnetic field from one wire is opposite to the other. This means the magnetic fields will cancel each other to some extent.

Multiple-Choice Questions

19.15 C (deflected toward the top of the page). The right hand rule and the Lorentz force law gives the direction of the force acting on a moving charge in a magnetic field.

19.19 A (zero). The force on each infinitesimal portion of the ring points radially, so the torque on the loop is equal to zero.

19.23 E (zero force). You can prove this to yourself either through the right hand rule and applying $\vec{F} = i\vec{l} \times \vec{B}$ or through symmetry.

Estimation/Numerical Questions

19.29 The time is equal to the distance traveled divided by the speed:

$$t = \frac{2\pi R}{v} = \frac{2\pi(0.1 \text{ m})}{\left(10^5 \frac{\text{m}}{\text{s}}\right)} \approx 6 \times 10^{-6} \text{ s} = 6 \ \mu\text{s}$$

Problems

19.33

SET UP

We are shown six scenarios of a positive charge moving with a velocity $\vec{v}$ in a magnetic field $\vec{B}$. We can use the Lorentz force law, $\vec{F} = q\vec{v} \times \vec{B}$, and the right hand rule to determine the direction of the magnetic force acting on the charge.

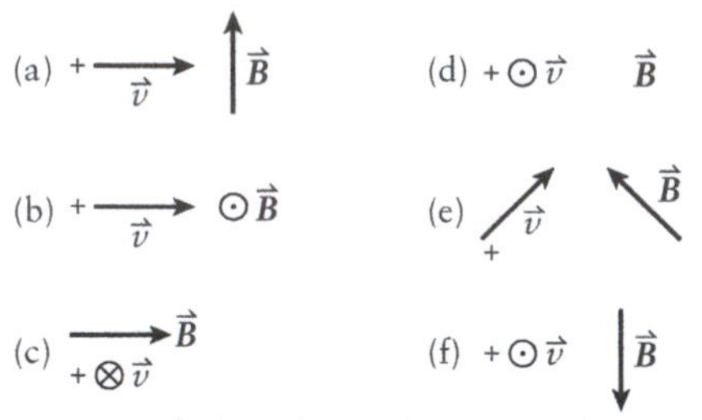

Figure 19-1 Problem 33

SOLVE

Part a) The force points out of the page.

Part b) The force points down.

Part c) The force points down.

Part d) The force points to the left.

Part e) The force points out of the page.

Part f) The force points to the right.

REFLECT
All of the answers would be reversed if the moving charge were negative.

19.37

SET UP
A proton ($q = 1.6 \times 10^{-19}$ C) travels with a speed of 18 m/s toward the top of the page through a uniform magnetic field of 2.0 T directed into the page. The magnitude and direction of the magnetic force acting on the proton is given by the Lorentz force law, $\vec{F} = q\vec{v} \times \vec{B}$, and the right hand rule.

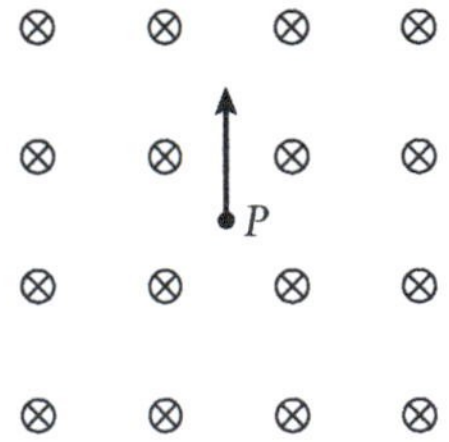

Figure 19-2 Problem 37

SOLVE
Magnitude:

$$F = qvB\sin(\varphi) = (1.6 \times 10^{-19}\text{ C})\left(18\frac{\text{m}}{\text{s}}\right)(2.0\text{ T})\sin(90°) = \boxed{5.8 \times 10^{-18}\text{ N}}$$

Direction: The force points to the $\boxed{\text{left}}$.

REFLECT

This is the maximum value of the force since $\vec{v}$ and $\vec{B}$ are perpendicular to one another.

Get Help: Interactive Example – Motion in a Magnetic Field
P'Cast 19.1 – Potassium Ion

19.41

SET UP

An electron ($q = -1.6 \times 10^{-19}$ C) travels with a speed of 10^7 m/s in the xy plane at an angle of 45 degrees above the $+x$-axis through a uniform magnetic field of 3.0 T directed towards $+y$. The magnitude and direction of the magnetic force acting on the proton is given by the Lorentz force law, $\vec{F} = q\vec{v} \times \vec{B}$, and the right hand rule.

SOLVE

Magnitude:

$$F = qvB\sin(\varphi) = (1.6 \times 10^{-19}\ \text{C})\left(10^7 \frac{\text{m}}{\text{s}}\right)(3.0\ \text{T})\sin(45°) = \boxed{3.4 \times 10^{-12}\ \text{N}}$$

Direction: The force points in the $\boxed{-z\ \text{direction}}$.

REFLECT

Remember to take the sign of the charge into account when assigning the direction of the magnetic force.

Get Help: Interactive Example – Motion in a Magnetic Field
P'Cast 19.1 – Potassium Ion

19.45

SET UP

A straight segment of wire that is 0.350 m long carries a current of 1.40 A in a uniform magnetic field. The segment makes an angle of 53 degrees with $\vec{B}$. The magnitude of the force acting on the segment is 0.200 N. We can calculate the magnitude of the magnetic field from the expression for the magnetic force on a current-carrying wire, $F = ILB\sin(\varphi)$.

SOLVE

$$F = ILB\sin(\varphi)$$

$$B = \frac{F}{IL\sin(\varphi)} = \frac{0.200\ \text{N}}{(1.40\ \text{A})(0.350\ \text{m})\sin(53°)} = \boxed{0.511\ \text{T}}$$

REFLECT

The force acts in a direction perpendicular to both the current flow and the magnetic field.

19.51

SET UP

A long wire of length l stretches along the y-axis and carries a current of $i = 1.0$ A in the $+y$ direction; we can represent the vector $i\vec{l}$ as $il\hat{y}$. This wire is in a uniform magnetic field

$\vec{B} = (0.10 \text{ T})\hat{x} - (0.20 \text{ T})\hat{y} + (0.30 \text{ T})\hat{z}$. The magnetic force on a current-carrying wire is $\vec{F} = i\vec{l} \times \vec{B}$. We can use the determinant form of the cross product to calculate the force on the wire per unit length, $\dfrac{\vec{F}}{l}$.

SOLVE

$$\vec{F} = i\vec{l} \times \vec{B} = (il\hat{y}) \times ((0.10 \text{ T})\hat{x} - (0.20 \text{ T})\hat{y} + (0.30 \text{ T})\hat{z}) = \begin{vmatrix} \hat{x} & \hat{y} & \hat{z} \\ 0 & il & 0 \\ 0.10 \text{ T} & -0.20 \text{ T} & 0.30 \text{ T} \end{vmatrix}$$

$$= (il(0.30 \text{ T}))\hat{x} - 0\hat{y} + (il(-0.10 \text{ T}))\hat{z}$$

$$\frac{\vec{F}}{l} = (i(0.30 \text{ T}))\hat{x} + (i(-0.10 \text{ T}))\hat{z} = ((1.0 \text{ A})(0.30 \text{ T}))\hat{x} + ((1.0 \text{ A})(-0.10 \text{ T}))\hat{z}$$

$$= \boxed{\left(0.30\frac{\text{N}}{\text{m}}\right)\hat{x} + \left(-0.10\frac{\text{N}}{\text{m}}\right)\hat{z}}$$

REFLECT

The force must be perpendicular to both $\vec{l}$ and $\vec{B}$; this means the dot products $\vec{l} \cdot \vec{F}$ and $\vec{B} \cdot \vec{F}$ must equal zero:

$$\vec{l} \cdot \vec{F} = (l\hat{y}) \cdot \left(\left(0.30\frac{\text{N}}{\text{m}}\right)l\hat{x} + \left(-0.10\frac{\text{N}}{\text{m}}\right)l\hat{z}\right) = 0$$

$$\vec{B} \cdot \vec{F} = ((0.10 \text{ T})\hat{x} - (0.20 \text{ T})\hat{y} + (0.30 \text{ T})\hat{z}) \cdot \left(\left(0.30\frac{\text{N}}{\text{m}}\right)l\hat{x} + \left(-0.10\frac{\text{N}}{\text{m}}\right)l\hat{z}\right)$$

$$= \left(0.03l\frac{\text{T} \cdot \text{N}}{\text{m}}\right) + \left(-0.03l\frac{\text{T} \cdot \text{N}}{\text{m}}\right) = 0$$

19.53

SET UP

A round loop of wire ($R = 0.10$ m) carries a current of $i_0 = 100$ A. The loop makes an angle of $\varphi = 30°$ with a magnetic field of $B = 0.244$ T. The angle is then changed to $\varphi = 10°$ and $\varphi = 50°$. The magnitude of the torque on the loop in each case is $\tau = i_0 AB \sin(\varphi)$.

SOLVE

$\varphi = 30°$:

$$\tau = i_0 AB \sin(\varphi) = (100 \text{ A})(\pi(0.10 \text{ m})^2)(0.244 \text{ T})\sin(30°) = \boxed{0.383 \text{ N} \cdot \text{m}}$$

$\varphi = 10°$:

$$\tau = i_0 AB \sin(\varphi) = (100 \text{ A})(\pi(0.10 \text{ m})^2)(0.244 \text{ T})\sin(10°) = \boxed{0.133 \text{ N} \cdot \text{m}}$$

$\varphi = 50°$:

$$\tau = i_0 AB \sin(\varphi) = (100 \text{ A})(\pi(0.10 \text{ m})^2)(0.244 \text{ T})\sin(50°) = \boxed{0.587 \text{ N} \cdot \text{m}}$$

REFLECT

The torque is a maximum when $\varphi = 90°$.

19.57

SET UP

Two finite wires carry antiparallel currents: wire 1 has a length $L_1 = 0.10$ m with a current $i_1 = 4$ A directed up, wire 2 has a length $L_2 = 0.08$ m with a current $i_2 = 3$ A directed down (see figure). Point P is located $d_1 = 0.07$ m to the right of wire 1 and $d_2 = 0.03$ m to the left of wire 2 along an axis cutting through the center of each wire. The magnetic field at P is the vector sum of the magnetic fields due to wires 1 and 2 at P. The field due to each wire points into the page at P, so the magnitude of the field at P is simply the sum of the magnitudes of the magnetic fields due to wires 1 and 2. We can use the Biot-Savart law to find an expression for the magnitude of the magnetic field due to a finite wire of length L and a distance d from the P in general before plugging in our values.

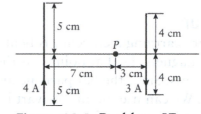

Figure 19-3 Problem 57

SOLVE

Magnetic field due to a finite wire:

$$B = \int_{-\frac{L}{2}}^{\frac{L}{2}} \frac{\mu_0}{4\pi} \frac{idl\sin(\varphi)}{r^2} = \frac{\mu_0 i}{4\pi} \int_{-\frac{L}{2}}^{\frac{L}{2}} \frac{dl\left(\dfrac{d}{r}\right)}{r^2} = \frac{\mu_0 id}{4\pi} \int_{-\frac{L}{2}}^{\frac{L}{2}} \frac{dl}{r^3} = \frac{\mu_0 id}{4\pi} \int_{-\frac{L}{2}}^{\frac{L}{2}} \frac{dl}{(d^2 + l^2)^{\frac{3}{2}}}$$

$$= \frac{\mu_0 id}{4\pi} \left[\frac{l}{d^2\sqrt{d^2 + l^2}} \right]_{-\frac{L}{2}}^{\frac{L}{2}} = \frac{\mu_0 id}{4\pi} \left[\frac{\left(\dfrac{L}{2}\right)}{d^2\sqrt{d^2 + \left(\dfrac{L}{2}\right)^2}} - \frac{\left(-\dfrac{L}{2}\right)}{d^2\sqrt{d^2 + \left(-\dfrac{L}{2}\right)^2}} \right]$$

$$= \frac{\mu_0 i}{4\pi d} \left[\frac{L}{\sqrt{d^2 + \left(\dfrac{L}{2}\right)^2}} \right]$$

Magnitude of the magnetic field at point P:

$$B_P = B_1 + B_2 = \frac{\mu_0 i_1}{4\pi d_1} \left[\frac{L_1}{\sqrt{d_1^2 + \left(\dfrac{L_1}{2}\right)^2}} \right] + \frac{\mu_0 i_2}{4\pi d_2} \left[\frac{L_1}{\sqrt{d_2^2 + \left(\dfrac{L_2}{2}\right)^2}} \right]$$

$$= \frac{\mu_0}{4\pi} \left[\frac{L_1 i_1}{d_1\sqrt{d_1^2 + \left(\dfrac{L_1}{2}\right)^2}} + \frac{L_2 i_2}{d_2\sqrt{d_2^2 + \left(\dfrac{L_2}{2}\right)^2}} \right]$$

$$= \frac{\left(4\pi \times 10^{-7}\dfrac{\text{T} \cdot \text{m}}{\text{A}}\right)}{4\pi} \left[\frac{(0.10 \text{ m})(4 \text{ A})}{(0.07 \text{ m})\sqrt{(0.07 \text{ m})^2 + (0.05 \text{ m})^2}} + \frac{(0.08 \text{ m})(3 \text{ A})}{(0.03 \text{ m})\sqrt{(0.03 \text{ m})^2 + (0.04 \text{ m})^2}} \right]$$

$$= 2.26 \times 10^{-5} \text{ T}$$

The net magnetic field at point P has a magnitude of 2.26×10^{-5} T and points into the page.

REFLECT

This is the same method outlined in Example 19-3 in the textbook.

19.61

SET UP

A wire carrying a current i is bent into two semicircular parts and two flat parts (see figure). The magnetic field at point C is the vector sum of the magnetic fields due to each of the four segments. Since the flat parts lie along the same axis as C, they do not contribute to the field at C. We can use the Biot-Savart law to find an expression for the magnitude of the magnetic field at the center of a semicircular wire for a general radius r; the radius will be constant for a given semicircular segment. The magnetic field at point C due to the closer, smaller semicircle of radius r points into the page (which we'll call $-z$), whereas the magnetic field at point C due to the farther, larger semicircle of radius R points out of the page (or $+z$).

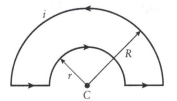

Figure 19-4 Problem 61

SOLVE

Magnitude of magnetic field due to a semicircular wire at its center:

$$B = \int_{\text{semicircle}} \frac{\mu_0}{4\pi} \frac{i|d\vec{l} \times \hat{r}|}{r^2} = \frac{\mu_0 i}{4\pi r^2} \int_{\text{semicircle}} dl = \frac{\mu_0 i}{4\pi r^2}(\pi r) = \frac{\mu_0 i}{2r}$$

Magnetic field at point C:

$$\vec{B} = -\left(\frac{\mu_0 i}{2r}\right)\hat{z} + \left(\frac{\mu_0 i}{2R}\right)\hat{z} = \frac{\mu_0 i}{2}\left(\frac{1}{R} - \frac{1}{r}\right)\hat{z} = \frac{\mu_0 i}{2}\left(\frac{r - R}{rR}\right)\hat{z} = \boxed{-\frac{\mu_0 i}{2}\left(\frac{R - r}{rR}\right)\hat{z}}$$

REFLECT

The magnetic field at point C points into the page because $R > r$. A tangent to a circle is always perpendicular to its radius so $|d\vec{l} \times \hat{r}| = |d\vec{l}||\hat{r}| = dl$. The integral of dl around the entire circle is equal to its circumference.

19.65

SET UP

A current of 100 A passes through a wire that is 5 m from a window. We can draw an Amperian loop with a radius $r = 5$ m around the wire and apply Ampere's law to find the magnitude of the magnetic field at the window. Once we know the magnitude of the field due to the power line, we can compare it to the magnitude of the Earth's magnetic field ($B_{\text{Earth}} = 0.5 \times 10^{-4}$ T).

SOLVE

$$\oint \vec{B} \cdot d\vec{l} = \mu_0 i_{\text{through}}$$

$$B \oint dl = B(2\pi r) = \mu_0 i$$

$$B = \frac{\mu_0 i}{2\pi r} = \frac{\left(4\pi \times 10^{-7}\dfrac{\text{T} \cdot \text{m}}{\text{A}}\right)(100 \text{ A})}{2\pi(5 \text{ m})} = \boxed{4 \times 10^{-6} \text{ T}}$$

$$\frac{B}{B_{\text{Earth}}} = \frac{4 \times 10^{-6} \text{ T}}{5 \times 10^{-5} \text{ T}} = 0.08$$

$$\boxed{B = 0.08 B_{\text{Earth}}}$$

REFLECT

Learning the algebraic expression for the magnitude of a magnetic field due to a straight, current-carrying wire will prove handy.

Get Help: Interactive Example – Magnetic Field from Current Loop

19.69

SET UP

A coaxial cable is made of a solid inner conductor of radius R_i and a concentric outer conducting shell of radius R_o. Looking down the cable, the inner conductor carries a current of i coming out of the page, which we'll define as positive current, and the outer conductor carries a current of i going into the page, which we'll define as negative current. There is an insulator in between these two conductors. In order to find the magnetic field in all space, we will use an Amperian loop of radius r and split the problem into three regions—1) $r < R_i$, 2) $R_i \leq r < R_o$, and 3) $r \geq R_o$—and apply Ampere's law. The Amperian loop in region 1 encloses a fraction of the current distributed throughout the solid inner conductor; assuming the current is uniformly distributed, the enclosed current is equal to the ratio of the cross-sectional area of the Amperian loop to the cross-sectional area of the inner conductor multiplied by the total current. The Amperian loop in region 2 encloses the entire inner conductor. Finally, the net current enclosed by the loop in region 3 is zero, which means the magnetic field in that region is also zero.

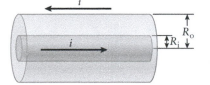

Figure 19-5 Problem 69

SOLVE

Looking at the cable head on:

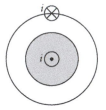

Figure 19-6 Problem 69

$r < R_i$:

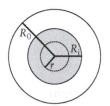

Figure 19-7 Problem 69

$$\oint \vec{B} \cdot d\vec{l} = \mu_0 i_{\text{through}}$$

$$B \oint dl = B(2\pi r) = \mu_0 \left(\frac{\pi r^2}{\pi R_i^2} \right) i$$

$$\boxed{B = \frac{\mu_0 i r}{2\pi R_i^2}}$$

$R_i \le r < R_o$:

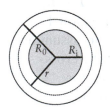

Figure 19-8 Problem 69

$$\oint \vec{B} \cdot d\vec{l} = \mu_0 i_{\text{through}}$$

$$B \oint dl = B(2\pi r) = \mu_0 i$$

$$\boxed{B = \frac{\mu_0 i}{2\pi r}}$$

$r \geq R_o$:

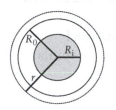

Figure 19-9 Problem 69

$$\oint \vec{B} \cdot d\vec{l} = \mu_0 i_{\text{through}} = \mu_0(i - i) = 0$$

$$\boxed{B = 0}$$

REFLECT
The field circulates counterclockwise in regions 1 and 2. As the Amperian loop increases in size in the region $r < R_i$, it encloses more and more current; therefore, the magnetic field should increase with r in this region. In the region $R_i \leq r < R_o$, we've fully enclosed the inner current i, so the magnetic field should decrease as r increases.

Get Help: Interactive Example – Coaxial Cylindrical Conductors
P'Cast 19.4 – Magnetic Field Due to a Coaxial Cable

19.73

SET UP
Wire 1 ($L_1 = 0.01$ m) carries a current of 2.00 A pointing north. Wire 2 carries a current of 3.60 A pointing south; wire 2 is $d = 1.40$ m to the right of wire 1. The magnitude of the force due to wire 2 on wire 1 is equal to $|i_1 \vec{L}_1 \times \vec{B}_2|$. As a reminder, the magnitude of a magnetic field a distance d away from a straight current-carrying wire is $\dfrac{\mu_0 i}{2\pi d}$. The direction of the force on wire 1 due to wire 2 is given by the right hand rule between the direction of i_1 and the direction of B_2. The magnetic field due to wire 2 points directly into the page at the location of wire 1, so the force will point to the left.

$i_1 = 2.00$ A, north $i_2 = 3.60$ A, south

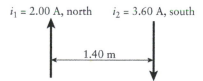

1.40 m

Figure 19-10 Problem 73

SOLVE
Magnitude:

$$F_{2\rightarrow1} = |i_1\vec{L}_1 \times \vec{B}_2| = i_1 L_1 B_2 = i_1 L_1\left(\frac{\mu_0 i_2}{2\pi d}\right) = \frac{\mu_0 i_1 i_2 L_1}{2\pi d}$$

$$= \frac{\left(4\pi \times 10^{-7}\dfrac{\text{T} \cdot \text{m}}{\text{A}}\right)(2.00 \text{ A})(3.60 \text{ A})(0.01 \text{ m})}{2\pi(1.40 \text{ m})} = 1.03 \times 10^{-8} \text{ N}$$

The force on wire 1 due to wire 2 has a magnitude of $\boxed{1.03 \times 10^{-8}\ \text{N and points to the left}}$.

REFLECT

In general, wires with antiparallel currents repel one another, while wires with parallel currents attract one another. You can prove this to yourself either by performing the similar calculations or by using Newton's third law.

Get Help: P'Cast 19.5 – Wires in a Computer

19.79

SET UP

An electron enters a region of crossed electric and magnetic fields known as a velocity selector. The electric field has a magnitude of E_{VS} = 90,000 V/m and points down (towards $-y$), and the magnetic field has a magnitude of B_{VS} = 0.0053 T and points into the page. In the velocity selector, the net force on the electron in the y direction is zero since the electron travels in a straight line. Therefore, the magnitude of the force due to the electric field must equal the magnitude of the force due to the magnetic field. Setting these equal gives us an expression for the speed of the electron as it exits the selector. Immediately after the selector, the electron enters a region with a magnetic field B_0 = 0.00242 T. The electron then undergoes uniform circular motion in a radius R = 0.04 m due to the magnetic force acting on it. We can solve for the mass of the electron by applying Newton's second law once again.

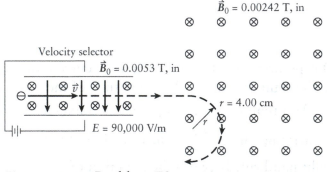

Figure 19-11 Problem 79

SOLVE

Free-body diagram of the electron in the velocity selector:

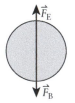

Figure 19-12 Problem 79

Speed of the electron after the velocity selector:

$$\sum F_y = F_E - F_B = ma_y = 0$$

$$qE_{VS} - qvB_{VS} = 0$$

$$v = \frac{E_{VS}}{B_{VS}} = \frac{\left(90{,}000\frac{V}{m}\right)}{0.0053 \text{ T}} = 1.698 \times 10^7 \frac{m}{s}$$

Free-body diagram of the electron as soon as it enters the spectrometer:

Figure 19-13 Problem 79

Mass of the electron:

$$\sum F_y = -F_B = ma_y = m\left(-\frac{v^2}{R}\right)$$

$$qvB_0 = \frac{mv^2}{R}$$

$$m = \frac{RqB_0}{v} = \frac{RqB_0B_{VS}}{E_{VS}}$$

$$= \frac{(0.040 \text{ m})(1.6 \times 10^{-19} \text{ C})(0.00242 \text{ T})(0.0053 \text{ T})}{\left(90{,}000\frac{V}{m}\right)} = \boxed{9.12 \times 10^{-31} \text{ kg}}$$

REFLECT

The accepted mass of the electron (to three significant figures) is 9.11×10^{-31} kg, so our value is reasonable.

> **Get Help:** Interactive Example – Motion in a Magnetic Field
> P'Cast 19.1 – Potassium Ion

19.85

SET UP

A thin, nonconducting ring of radius R and total charge Q lies with its center at the origin of xy plane. The ring is spinning at an angular speed of ω about its center. This spinning charge will generate a magnetic field at point P, which is a vertical distance Z above the xy plane and a distance $r = \sqrt{Z^2 + R^2}$ from the ring, that we can determine through the Biot-Savart law. Since the ring is centered about the z-axis and is a constant distance R from the axis, we can convert the integral from one over dl to one over $d\theta$ through the definition of the arc length. The current is equal to the change in charge divided by the change in time. For the spinning ring, it completes one full revolution in $\frac{2\pi}{\omega}$. In this time period the entire charge of Q has gone around, which means the current is $i = \frac{Q}{\left(\frac{2\pi}{\omega}\right)}$. The magnetic field at point P can only point along the +z-axis due to symmetry, so we need to multiply our result by $\cos(\phi)$, where ϕ is the angle between the magnetic field and the z-axis.

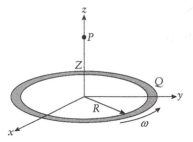

Figure 19-14 Problem 85

SOLVE
Current:

$$i = \frac{\Delta Q}{\Delta T} = \frac{Q}{\left(\dfrac{2\pi}{\omega}\right)} = \frac{Q\omega}{2\pi}$$

Magnitude of magnetic field at P:

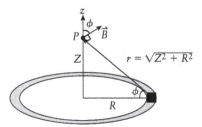

Figure 19-15 Problem 85

$$B_P = \int \frac{\mu_0 i |d\vec{l} \times \hat{r}|}{4\pi r^2} = \frac{\mu_0 i}{4\pi} \int \frac{dl}{r^2} = \frac{\mu_0 i}{4\pi} \int_0^{2\pi} \frac{R\,d\theta}{(Z^2 + R^2)} = \frac{\mu_0 i R}{4\pi (Z^2 + R^2)} \int_0^{2\pi} d\theta = \frac{\mu_0 i R}{4\pi (Z^2 + R^2)} [\theta]_0^{2\pi}$$

$$= \frac{\mu_0 i R}{4\pi (Z^2 + R^2)} [2\pi] = \frac{\mu_0 \left(\dfrac{Q\omega}{2\pi}\right) R}{2(Z^2 + R^2)} = \frac{\mu_0 Q\omega R}{4\pi (Z^2 + R^2)}$$

z component of magnetic field at P:

$$B_{P,z} = B_P \cos(\phi) = \left(\frac{\mu_0 Q\omega R}{4\pi (Z^2 + R^2)}\right)\left(\frac{R}{\sqrt{Z^2 + R^2}}\right) = \frac{\mu_0 Q\omega R^2}{4\pi (Z^2 + R^2)^{\frac{3}{2}}}$$

The magnetic field at point P has a magnitude of $\boxed{\dfrac{\mu_0 Q\omega R^2}{4\pi (Z^2 + R^2)^{\frac{3}{2}}}$ and points towards $+z}$.

REFLECT
The right-hand rule for a loop of current shows that the magnetic field generated by the current points towards +z within the loop, which is consistent with our answer.

19.91

SET UP

Two straight conducting rods—rod 1 has a resistance $R_1 = 0.5\ \Omega$, rod 2 has a resistance $R_2 = 2.5\ \Omega$—that have the same mass $m = 25 \times 10^{-3}$ kg and length $l = 1.0$ m are connected in series by a resistor ($R = 17\ \Omega$) and an external voltage source with potential difference ε_0. We are told that rod 1 "floats" a distance $d = 0.85 \times 10^{-3}$ m above rod 2, which means the net force acting on rod 1 is zero. The forces acting on rod 1 are the force due to the magnetic field generated by rod 2 pointing up and the force due to gravity pointing down. In order to find the magnetic field generated by rod 2, we first need to find the current in the circuit. The two rods and the resistor are in series, which means the current is constant throughout the circuit. The current is equal to the potential difference ε_0 divided by the equivalent resistance of the circuit. The magnitude of the magnetic field due to rod 2 at rod 1 is $B_2 = \dfrac{\mu_0 i}{2\pi d}$ and the magnitude of the magnetic force acting on rod 2 is ilB_2. Setting this expression equal to the force of gravity, we can calculate ε_0.

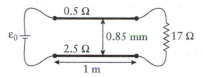

Figure 19-16 Problem 91

SOLVE

Equivalent resistance:

$$R_{equiv} = R_1 + R_2 + R = (0.5\ \Omega) + (2.5\ \Omega) + (17\ \Omega) = 20\ \Omega$$

Current in the circuit:

$$\varepsilon_0 = iR_{equiv}$$

$$i = \frac{\varepsilon_0}{R_{equiv}}$$

Magnetic field due to rod 2 at the location of rod 1:

$$B_2 = \frac{\mu_0 i}{2\pi d} = \frac{\mu_0 \left(\dfrac{\varepsilon_0}{R}\right)}{2\pi d} = \frac{\mu_0 \varepsilon_0}{2\pi d R_{equiv}}$$

Free-body diagram for rod 2:

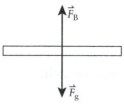

Figure 19-17 Problem 91

Newton's second law for rod 1:

$$\sum F_y = F_B - F_g = ma_y = 0$$

$$F_B = F_g$$

$$ilB_2 = mg$$

$$\left(\frac{\varepsilon_0}{R_{equiv}}\right)l\left(\frac{\mu_0\varepsilon_0}{2\pi dR_{equiv}}\right) = mg$$

$$\varepsilon_0 = \sqrt{\frac{2\pi dR_{equiv}^2 mg}{\mu_0 l}} = \sqrt{\frac{2\pi(0.85 \times 10^{-3}\ \text{m})(20\ \Omega)^2(25 \times 10^{-3}\ \text{kg})\left(9.8\frac{\text{m}}{\text{s}^2}\right)}{\left(4\pi \times 10^{-7}\frac{\text{T}\cdot\text{m}}{\text{A}}\right)(1.0\ \text{m})}} = \boxed{650\ \text{V}}$$

REFLECT

In the space directly above rod 2, its magnetic field points into the page. The current through rod 1 is moving towards the right. The cross product $l_1 \times B_2$ points up as expected since the net force on rod 1 is zero.

Get Help: P'Cast 19.5 – Wires in a Computer

19.95

SET UP

A high voltage power line is located 5.0 m in the air and 12 m horizontally from your house. The wire carries a current of 100 A. We can use the expression $B = \frac{\mu_0 i}{2\pi r}$, where r is the straight-line distance from the wire to your house, to calculate the magnitude of the magnetic field caused by the power line. Once we know the magnitude of the field due to the power line, we can compare it to the magnitude of the Earth's magnetic field ($B_{Earth} = 0.5 \times 10^{-4}\ \text{T}$) to determine if the power line is likely to affect your health.

SOLVE

$$B = \frac{\mu_0 i}{2\pi r} = \frac{\left(4\pi \times 10^{-7}\frac{\text{T}\cdot\text{m}}{\text{A}}\right)(100\ \text{A})}{2\pi\sqrt{(5.0)^2 + (12\ \text{m})^2}} = \boxed{1.5 \times 10^{-6}\ \text{T}}$$

$$\frac{B}{B_{Earth}} = \frac{1.5 \times 10^{-6}\ \text{T}}{0.5 \times 10^{-4}\ \text{T}} = 0.031$$

$$\boxed{B = 0.031 B_{Earth}}$$

Since the magnetic field from the wires is so much smaller than the Earth's magnetic field, there should be $\boxed{\text{little or no cause for concern}}$.

REFLECT

Even if we were 1 m from the power line, the magnitude of the magnetic field would still only be 40% of the strength of the Earth's field.

Chapter 20
Magnetic Induction

Conceptual Questions

20.1 The falling bar magnet induces a current in the wall of the pipe. In accordance with Lenz's law, the direction of this induced current is such that it produces a magnetic field that exerts a force on the magnet opposing its motion. The speed of the magnet therefore increases more slowly than in free-fall until it reaches a terminal speed at which the magnetic and gravitational forces are equal and opposite. After this point, the magnet continues to fall, but at a constant speed.

20.7 Inductance is flux divided by current. Since flux is proportional to current, the current terms cancel, leaving only geometric properties.

20.11 Part a) Assuming the number of coils remains constant, the self-inductance drops by a factor of 2.

Part b) Assuming the length of the solenoid remains constant, the self-inductance is unchanged.

Multiple-Choice Questions

20.15 D (d). The loop is starting to exit the region with the magnetic field.

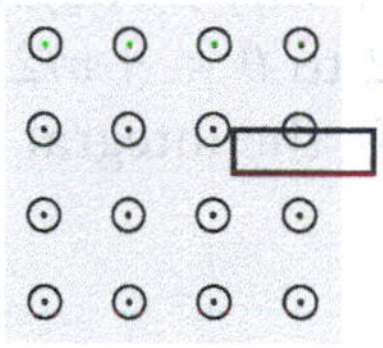

(d)

Figure 20-1 Problem 15

20.17 C (antiparallel to i_a). The induced current in loop b will flow in such a way as to generate a magnetic field to counteract the increase in flux due to loop a.

20.21 C (one-quarter of the total energy).

$$U_E = \frac{1}{2C}q_{max}^2$$

$$\frac{U_{half\,max}}{U_{max}} = \frac{\left(\frac{1}{2C}\left(\frac{q_{max}}{2}\right)^2\right)}{\left(\frac{1}{2C}q_{max}^2\right)} = \frac{1}{4}$$

Estimation/Numerical Questions

20.27 Clearly it was an extremely important achievement for Faraday to conceive of induction (at any time in history). However, it was critical that his accomplishment occur at the approximate time that it did (~1825) because it took several decades for other scientists to fully study and perfect his ideas. Without this fundamental work, it is questionable whether Maxwell would have been able to synthesize electricity and magnetism with optics as he did in 1865. Without Maxwell's equations (and the field theory that followed), the course of scientific history would have been dramatically altered and the entire quantum revolution of the earth 20th century might never have occurred!

Problems

20.31

SET UP

A ring of radius $R = 0.0800$ m is lying the xy plane centered about the origin. The ring is in the presence of a magnetic field, $\vec{B} = (6.00\text{ T}) \cos(\theta)\hat{z}$, where θ is measured in the xy plane with respect to the positive x-axis. The magnitude of the magnetic flux through the ring is

$\Phi_B = \int \vec{B} \cdot d\vec{A}$, where $d\vec{A}$ is the infinitesimal area vector for a circle. The area vector points normal to the surface, which is towards $+z$ in this case. Since the two vectors are parallel, the dot product $\vec{B} \cdot d\vec{A} = BdA$. The infinitesimal area dA in cylindrical coordinates is $rdrd\theta$. To calculate the flux through the ring, the limits for the integral over dr are $r = 0$ to $r = R$ and for the integral over $d\theta$ are $\theta = 0$ to $\theta = 2\pi$. If we're interested in the flux only through the part of the circle where $x > 0$, the integral over $d\theta$ will be from $\theta = -\pi/2$ to $\theta = +\pi/2$. If we're interested in the flux only through the part of the circle where $y > 0$, the integral over $d\theta$ will be from $\theta = 0$ to $\theta = \pi$.

SOLVE

Part a)

$$\Phi_B = \int \vec{B} \cdot d\vec{A} = \int ((6.00\text{ T}) \cos(\theta)\hat{z}) \cdot (dA\hat{z}) = (6.00\text{ T}) \iint \cos(\theta)(rdrd\theta)$$

$$= (6.00\text{ T})\int_0^R rdr \int_0^{2\pi} \cos(\theta)d\theta = (6.00\text{ T})\left[\frac{1}{2}r^2\right]_0^R [\sin(\theta)]_0^{2\pi} = (6.00\text{ T})\left[\frac{1}{2}r^2\right]_0^R [0] = \boxed{0}$$

Part b)

$$\Phi_B = \int \vec{B} \cdot d\vec{A} = (6.00\text{ T})\int_0^R rdr \int_{-\frac{\pi}{2}}^{\frac{\pi}{2}} \cos(\theta)d\theta = (6.00\text{ T})\left[\frac{1}{2}r^2\right]_0^R [\sin(\theta)]_{-\frac{\pi}{2}}^{\frac{\pi}{2}}$$

$$= (6.00\text{ T})\left[\frac{1}{2}R^2\right]\left[\sin\left(\frac{\pi}{2}\right) - \sin\left(-\frac{\pi}{2}\right)\right] = \frac{(6.00\text{ T})}{2}(0.0800\text{ m})^2[2] = \boxed{0.384\text{ Wb}}$$

Part c)

$$\Phi_B = \int \vec{B} \cdot d\vec{A} = (6.00 \text{ T}) \int_0^R r\,dr \int_0^\pi \cos(\theta)\,d\theta = (6.00 \text{ T}) \left[\frac{1}{2} r^2 \right]_0^R [\sin(\theta)]_0^\pi$$

$$= (6.00 \text{ T}) \left[\frac{1}{2} R^2 \right] [\sin(\pi) - \sin(0)] = \boxed{0}$$

REFLECT

We could have also used symmetry to solve parts a and c. Cosine is an even function and positive in the first and fourth quadrants, where $x > 0$. The flux will be positive for values where $x > 0$ and equal in magnitude but negative for values of $x < 0$. Therefore, an integral over the entire ring or over the semicircle where $y > 0$ will be exactly zero.

20.35

SET UP

A bar magnet, leading with its south pole, is moved at a constant velocity through a wire loop. We'll assume the magnet starts infinitely far to the left of the loop at $t = 0$. As the bar magnet passes through the loop, the magnetic flux is constantly changing, which induces a voltage in the loop according to Faraday's and Lenz's law. In order to draw a qualitative sketch of the induced voltage in the loop, we will define the area vector of the loop to point towards the initial location of the magnet (to the left in the above figure); this defines our sense of positive induced voltage. The magnetic field pointing to the left is getting stronger as the magnet nears the loop before t_1. Since this corresponds to an increasing positive flux, this will generate an increasingly negative induced voltage until it reaches a minimum. Similarly, after the magnet leaves the loop, the magnetic field pointing to the left is getting weaker. This corresponds to a decreasing positive flux, which results in an increasingly positive induced voltage until it reaches a maximum. The flux is changing the most at t_2, so this corresponds to a maximum in the plot of the magnetic flux versus time. A maximum in the flux corresponds to an induced voltage of zero since the induced current will change direction at that moment.

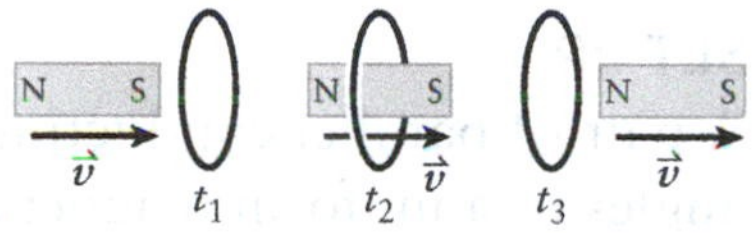

Figure 20-2 Problem 35

SOLVE

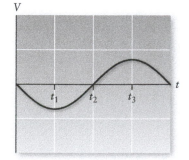

Figure 20-3 Problem 35

REFLECT

The induced voltage is related to the slope of the magnetic flux as a function of time.

20.37

SET UP

An electromagnetic generator consists of a wound coil ($N = 100$) that has an area of $A = 400$ cm^2. The coil rotates at $\omega = 60\frac{\text{rev}}{\text{s}}$ in a magnetic field with strength $B = 0.25$ T. The maximum induced potential is given by $\varepsilon_{\text{max}} = NBA(2\pi\omega)$, where ω is in rev/s.

SOLVE

$$\varepsilon_{\text{max}} = NBA(2\pi\omega) = (100)(0.25 \text{ T})\left(400 \text{ cm}^2 \times \left(\frac{1 \text{ m}}{100 \text{ cm}}\right)^2\right)(2\pi)\left(60\frac{\text{rev}}{\text{s}}\right) = \boxed{377 \text{ V}}$$

REFLECT

This is the maximum value; the actual potential oscillates between $+377$ V and -377 V.

20.41

SET UP

A pair of parallel conducting rails are a distance $L = 0.12$ m apart and situated at right angles to a uniform magnetic field $B = 0.8$ T pointed into the page. A resistor ($R = 15$ Ω) is connected across the rails. A conducting bar is placed on top of the rails and is moved at a constant speed of $v = 2$ m/s to the right. The bar creates a closed loop with a portion of each rail and the resistor. The area of the loop is equal to Lx, where x is the horizontal distance between the resistor and the bar; we'll assume the area vector also points into the page. The area of this closed loop increases as the bar moves to the right, which means the magnetic flux is also increasing. We can use Faraday's law and Ohm's law to calculate the magnitude of the current flowing through the resistor. Since the magnetic flux into the page is increasing, current will be induced in such a way as to counteract this increase according to Lenz's law. From the right hand rule, this corresponds to a counterclockwise current through the loop, so the current through the bar points up. The bar will experience a magnetic force due to the external magnetic field interacting with the current. The current is perpendicular to the field. In this case, the magnitude of the force is given by $F = iLB$, and the direction is given by the right hand rule for cross products.

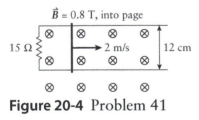

Figure 20-4 Problem 41

SOLVE

Part a)

Induced potential:

$$\varepsilon = -\frac{d\Phi_B}{dt} = -\frac{d}{dt}[\vec{B} \cdot \vec{A}] = -\frac{d}{dt}[BA] = -B\frac{d}{dt}[A] = -B\frac{d}{dt}[Lx] = -BL\frac{d}{dt}[x]$$

$$= -BLv = -(0.8 \text{ T})(0.12 \text{ m})\left(2\frac{\text{m}}{\text{s}}\right) = -0.192 \text{ V}$$

Current:

$$\varepsilon = iR$$

$$i = \frac{\varepsilon}{R} = \frac{|-0.192 \text{ V}|}{15\ \Omega} = \boxed{0.0128 \text{ A}}$$

Part b) The current in the bar points $\boxed{\text{up}}$.

Part c)

$$\vec{F} = i\vec{L} \times \vec{B}$$

$$F = iLB = (0.0128 \text{ A})(0.12 \text{ m})(0.8 \text{ T}) = 0.00123 \text{ N}$$

The magnetic force on the bar has a magnitude of $\boxed{0.00123 \text{ N and points to the left}}$.

REFLECT

By defining our area vector to point down, we've implicitly chosen a clockwise current to be positive. The negative sign in our induced potential means we have a "negative" current in the loop, *i.e.*, one that flows counterclockwise.

20.45

SET UP

A radio antenna is made from a solenoid of length $l = 0.030$ m and cross-sectional area $A = 0.50$ cm^2. The solenoid, which is filled with air, consists of $N = 300$ turns copper wire. The inductance of a solenoid is $L = \dfrac{\mu_0 N^2 A}{l}$.

SOLVE

$$L = \frac{\mu_0 N^2 A}{l} = \frac{\left(4\pi \times 10^{-7}\frac{\text{H}}{\text{m}}\right)(300)^2\left(0.50 \text{ cm}^2 \times \left(\frac{1 \text{ m}}{100 \text{ cm}}\right)^2\right)}{(0.030 \text{ m})}$$

$$= \boxed{1.88 \times 10^{-4} \text{ H} = 0.188 \text{ mH}}$$

REFLECT

The inductance of a solenoid only depends on its dimensions and a physical constant, so the fact that the wire is copper does not affect the inductance.

20.49

SET UP

An LC circuit is made up of an inductor ($L = 14.4 \times 10^{-3}$ H) and a fully charged capacitor ($C = 225 \times 10^{-6}$ F, $Q_0 = 300\ \mu$C). The circuit is completed at $t = 0$. The charge on the capacitor as a function of time is $q(t) = Q_0 \cos(\omega_0 t + \phi)$, where the natural frequency $\omega_0 = \sqrt{\dfrac{1}{LC}}$. We can determine the phase angle ϕ from the initial conditions of the system.

Once we have the expression for the charge as a function of time, we can solve for the charge at $t = 7.5 \times 10^{-3}$ s.

SOLVE
Natural frequency:

$$\omega_0 = \sqrt{\frac{1}{LC}} = \sqrt{\frac{1}{(14.4 \times 10^{-3}\ \text{H})(225 \times 10^{-6}\ \text{F})}} = 555.6\frac{\text{rad}}{\text{s}}$$

Charge as a function of time:

$$q(t) = Q_0 \cos(\omega_0 t + \phi)$$

$$q(0) = Q_0 = Q_0 \cos(\phi)$$

$$\phi = \arccos(1) = 0$$

$$q(t) = Q_0 \cos(\omega_0 t)$$

Charge at $t = 7.5 \times 10^{-3}$ s:

$$q(7.5 \times 10^{-3}\ \text{s}) = (300\ \mu\text{C})\cos\left(\left(555.6\frac{\text{rad}}{\text{s}}\right)(7.5 \times 10^{-3}\ \text{s})\right) = \boxed{-156\ \mu\text{C}}$$

REFLECT
The negative sign in our calculation for the charge at $t = 7.5 \times 10^{-3}$ s means the polarity of the capacitor has flipped relative to its initial setup—the positive plate is now negative and *vice versa*.

20.53

SET UP
An LC circuit is made up of an inductor ($L = 400 \times 10^{-3}$ H) and a capacitor ($C = 100 \times 10^{-6}$ F). In an LC circuit, both the current and charge are time dependent with the current 90 degrees out of phase with the charge. The maximum current in this particular circuit occurs at $t = 0$, which means the charge on the capacitor is 0. Since the charge and current differ in phase by 90 degrees, it will take one-quarter of a period for the charge to reach its maximum value. The period of the oscillation is equal to $\frac{2\pi}{\omega_0}$, where $\omega_0 = \sqrt{\frac{1}{LC}}$.

SOLVE
Natural frequency:

$$\omega = \sqrt{\frac{1}{LC}} = \sqrt{\frac{1}{(400 \times 10^{-3}\ \text{H})(100 \times 10^{-6}\ \text{F})}} = 158\frac{\text{rad}}{\text{s}}$$

Period of oscillation:

$$T = \frac{2\pi}{\omega} = \frac{2\pi}{\left(158\frac{\text{rad}}{\text{s}}\right)} = 0.0398\ \text{s}$$

One-quarter period:

$$\frac{T}{4} = \frac{0.0398 \text{ s}}{4} = \boxed{0.01 \text{ s}}$$

REFLECT

We could have also solved this using calculus. The extrema of a function are found by setting the derivative equal to zero and solving. We're interested in the maximum charge; the first derivative of the charge with respect to time is the current. The current as a function of time in this LC circuit is $i(t) = i_{max}\cos(\omega_0 t)$. Setting this equal to zero and solving for t, we find that

$$t = \frac{\arccos(0)}{\omega_0} = \frac{\left(\dfrac{\pi}{2}\right)}{\omega_0} = \frac{2\pi}{4\omega_0} = \frac{T}{4}.$$

20.59

SET UP

An LR circuit is made up of an inductor ($L = 22 \times 10^{-3}$ H) and a resistor ($R = 360 \ \Omega$). The time constant for an LR circuit is $\tau = \dfrac{L}{R}$.

SOLVE

$$\tau = \frac{L}{R} = \frac{22 \times 10^{-3} \text{ H}}{360 \ \Omega} = \boxed{6.11 \times 10^{-5} \text{ s}}$$

REFLECT

The current in an LR circuit exponentially decays or grows, whereas the current in an LC circuit oscillates in time.

20.63

SET UP

An LR circuit is made up of an inductor ($L = 12$ H) and a resistor ($R = 3.0 \ \Omega$). The current as a function of time in an LR circuit is given by $i(t) = \dfrac{V}{R}\left(1 - e^{-\frac{t}{\tau}}\right) = i_{max}\left(1 - e^{-\frac{t}{\tau}}\right)$, where the time constant $\tau = \dfrac{L}{R}$. We can rearrange this equation to calculate the time when $i(t) = 0.5 i_{max}$.

SOLVE

$$i(t) = \frac{V}{R}\left(1 - e^{-\frac{R}{L}t}\right) = i_{max}\left(1 - e^{-\frac{R}{L}t}\right)$$

$$t = -\left(\frac{L}{R}\right)\ln\left(1 - \frac{i(t)}{i_{max}}\right) = -\left(\frac{L}{R}\right)\ln\left(1 - \frac{0.5 i_{max}}{i_{max}}\right) = -\left(\frac{12 \text{ H}}{3.0 \ \Omega}\right)\ln(0.5) = \boxed{2.8 \text{ s}}$$

REFLECT

The time constant for this circuit is 4.0 s. After 4.0 s, the current is 63% of its maximum. Since we're interested in the time it takes to reach 50% of its maximum, our answer should be less than one time constant.

20.67

SET UP

An LR circuit is made up of a battery ($\varepsilon = 40$ V), an inductor ($L = 0.120$ H), and a resistor ($R = 8\ \Omega$). At some arbitrary moment in time, the current through the resistor is measured to be $i = 6.00$ A to the left. We can use Kirchhoff's loop rule to calculate the rate of change of the current, the potential difference $V_a - V_b$ at that moment, and the conditions under which the current is increasing. The induced potential difference across an inductor has a magnitude of $L\dfrac{di}{dt}$; the potential difference across a resistor has a magnitude of Ri.

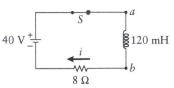

Figure 20-5 Problem 67

SOLVE

Part a)

Kirchhoff's loop rule:

$$\varepsilon - V_L - V_R = 0$$

$$\varepsilon - \left(L\frac{di}{dt}\right) - Ri = 0$$

$$\frac{di}{dt} = \frac{\varepsilon - Ri}{L} = \frac{(40\text{ V}) - (8\ \Omega)(6.00\text{ A})}{0.120\text{ H}} = \boxed{-66.7\frac{\text{A}}{\text{s}}}$$

Part b) The current is $\boxed{\text{decreasing}}$ because $\dfrac{di}{dt} < 0.$

Part c)

$$V_L = \varepsilon - V_R = \varepsilon - Ri = (40\text{ V}) - (8\ \Omega)(6.00\text{ A}) = \boxed{-8.00\text{ V}}$$

Part d) If $i < 5$ A, then $V_b < V_a$, which would mean the potential difference across the inductor is opposing that of the battery, which means the current is increasing.

REFLECT

At $i = 5$ A, the rate of change of the current is equal to zero. For $i < 5$ A the current is increasing, for $i > 5$ A the current is decreasing; this means 5 A is the maximum value of the current in the circuit.

20.71

SET UP

A magnetic field of $B = 0.45 \times 10^{-4}$ T is directed straight down, perpendicular to the plane of a circular coil of wire. The wire is made up of $N = 250$ turns and has an initial radius of $r_1 = $

0.20 m. The radius of the circle is increased to $r_2 = 0.30$ m in a period of $\Delta t = 15 \times 10^{-3}$ s. For simplicity, we'll assume the rate at which the radius changes is constant, the number of coils remains constant, and that the resistance of the coil also remains constant at $R = 25\ \Omega$ throughout the stretch. We can use Faraday's law of induction, $\varepsilon = N\dfrac{d\Phi_B}{dt}$, to calculate the voltage induced across the coil. Once we have the induced voltage across the coil, we can use Ohm's law to calculate the inducted current in the coil. Finally, the flux directed downward through the coil is increasing when the area of the loop is increasing; the induced magnetic field should point upward, so as to counteract this increase in flux. Therefore, the induced current will flow counterclockwise (when viewed from above), as given by the right-hand rule.

SOLVE

Part a)

$$\varepsilon = N\frac{d\Phi_B}{dt} = N\frac{d}{dt}[\vec{B}\cdot\vec{A}] = NB\frac{dA}{dt} = NB\left(\frac{A_f - A_i}{\Delta t}\right) = NB\pi\left(\frac{r_f^2 - r_i^2}{\Delta t}\right)$$

$$= (250)(0.45 \times 10^{-4}\ \text{T})\pi\left(\frac{(0.30\ \text{m})^2 - (0.20\ \text{m})^2}{15 \times 10^{-3}\ \text{s}}\right) = \boxed{0.118\ \text{V}}$$

Part b)

$$\varepsilon = iR$$

$$i = \frac{\varepsilon}{R} = \frac{0.118\ \text{V}}{25\ \Omega} = \boxed{0.00471\ \text{A}}$$

Part c) The induced current in the loop is $\boxed{\text{counterclockwise}}$ when viewed from above.

REFLECT
Our assumption that the resistance remained constant made the problem much easier to solve, but was an oversimplification. The resistance of the wire would increase linearly as we increase the radius of the coil because we've changed the dimensions of the coil.

20.77

SET UP
A pot contains a round metal coil that is 0.120 m in diameter and has a resistance $R = 22.5 \times 10^{-3}\ \Omega$. The surface of the stove produces a uniform vertical magnetic field that oscillates sinusoidally in time with amplitude $B_0 = 0.850$ T and frequency 60.0 Hz. The time-changing magnetic field induces a time-changing potential in the metal coil according to Faraday's law. Since the magnetic field and the area vector are parallel to one another, the dot product is equal to the product of their magnitudes. The average rate of heat produced in a resistive wire is equal to the time average of the potential squared divided by the resistance, $\langle P \rangle = \dfrac{\langle \varepsilon^2 \rangle}{R}$. Assuming all of this heat generated by the coil is absorbed by $m = 0.500$ kg of water $\left(c = 4186\dfrac{\text{J}}{\text{kg}\cdot\text{K}}\right)$, we can calculate the time required to increase the temperature of the water by 30 K.

SOLVE

Part a)

Induced potential:

$$\varepsilon = -\frac{d\Phi_B}{dt} = -\frac{d}{dt}[\vec{B}\cdot\vec{A}] = -\frac{d}{dt}[(B_0\cos(\omega t))A] = -B_0 A\frac{d}{dt}[\cos(\omega t)] = B_0 A\omega\,\sin(\omega t)$$

Average rate of heat produced:

$$\langle P\rangle = \frac{\langle\varepsilon^2\rangle}{R} = \frac{\langle(B_0 A\omega\,\sin(\omega t))^2\rangle}{R} = \frac{(B_0 A\omega)^2\langle\sin^2(\omega t)\rangle}{R} = \frac{(B_0 A\omega)^2\left(\frac{1}{2}\right)}{R}$$

$$= \frac{\left((0.850\text{ T})\left(\pi\left(\frac{0.120\text{ m}}{2}\right)^2\right)(2\pi(60.0\text{ Hz}))\right)^2}{2(22.5\times10^{-3}\ \Omega)} = \boxed{292\text{ W}}$$

Part b)

$$Q = Pt = mc\Delta T$$

$$t = \frac{mc\Delta T}{P} = \frac{(0.500\text{ kg})\left(4186\frac{\text{J}}{\text{kg}\cdot\text{K}}\right)(30\text{ K})}{292\text{ W}} = 215\text{ s}\times\frac{1\text{ min}}{60\text{ s}} = \boxed{3.59\text{ min}}$$

REFLECT

It would take about 9.5 min for the water to start boiling, which seems on the long side of reasonable. When calculating the time averages, recall that the average of a constant is equal to the constant and the average over one period of $\sin^2(\omega t) = \frac{1}{2}$.

20.79

SET UP

A tightly wound solenoid ($N_{\text{solenoid}} = 1500$, $l_{\text{solenoid}} = 0.180$ m, $d_{\text{solenoid}} = 0.0200$ m) carrying a current i_{solenoid} is surrounded by a circular coil ($N_{\text{coil}} = 20$, $d_{\text{coil}} = 0.0300$ m). The circular coil is connected across a resistor of very high resistance, which means the current through the coil is essentially zero for all time. The current in the solenoid is changing at a rate of $\frac{di_{\text{solenoid}}}{dt} = 100\frac{\text{A}}{\text{s}}$. The induced potential in the coil is equal to the sum of the induced potential due to the self-inductance of the coil and the induced potential due to the mutual inductance from the solenoid. Since the current in the coil remains essentially constant, the self-inductance term goes to zero. The mutual inductance for the system is given by $M = \frac{N_{\text{coil}}\Phi_{B,\text{coil}}}{i_{\text{solenoid}}}$. The magnetic flux through the coil arises from the magnetic field produced by the solenoid, $B_{\text{solenoid}} = \mu_0\frac{N_{\text{solenoid}}}{l_{\text{solenoid}}}i_{\text{solenoid}}$. This magnetic field is only produced in the region inside the solenoid, which is smaller than the cross-sectional area of the coil; therefore, when calculating the flux, the effective area is the cross-sectional area of the solenoid, not the coil.

The problem would be a bit more difficult if we had to calculate the magnetic flux through the solenoid due to the coil since the magnitude and direction of the magnetic field due to the coil changes along the radius and length of the solenoid. The easiest method would involve invoking the mutual inductance of the system and using our result from part a.

Figure 20-6 Problem 79

SOLVE

Part a)

$$\varepsilon_{coil} = -L_{coil}\frac{di_{coil}}{dt} - M\frac{di_{solenoid}}{dt} \approx 0 - M\frac{di_{solenoid}}{dt} = -\left(\frac{N_{coil}\Phi_{B,\,coil}}{i_{solenoid}}\right)\frac{di_{solenoid}}{dt}$$

$$= -\left(\frac{N_{coil}}{i_{solenoid}}\right)(B_{solenoid}A_{solenoid})\frac{di_{solenoid}}{dt}$$

$$= -\left(\frac{N_{coil}}{i_{solenoid}}\right)\left(\mu_0\frac{N_{solenoid}}{l_{solenoid}}i_{solenoid}\right)\left(\pi\left(\frac{d_{solenoid}}{2}\right)^2\right)\frac{di_{solenoid}}{dt}$$

$$= -\frac{\mu_0\pi}{4}\left(\frac{N_{solenoid}N_{coil}d_{solenoid}^2}{l_{solenoid}}\right)\left(\frac{di_{solenoid}}{dt}\right)$$

$$|\varepsilon_{coil}| = \frac{\left(4\pi \times 10^{-7}\frac{H}{m}\right)\pi}{4}\left(\frac{(1500)(20)(0.0200\text{ m})^2}{0.180\text{ m}}\right)\left(100\frac{A}{s}\right)$$

$$= \boxed{0.00658\text{ V} = 6.58\text{ mV}}$$

Part b) In the new situation, it is much more difficult to calculate the flux through the solenoid from the coil because the magnitude and direction of the magnetic field change along the length and radius of the solenoid, respectively. The problem is not impossible, though. From the existing problem, we can calculate the mutual inductance, M. Once we have that, the problem is reduced to $\varepsilon_{solenoid} = M\frac{di_{coil}}{dt}$.

REFLECT

A few mV seems reasonable for an induced potential. In general, if you need a quantity that was not provided in the problem statement, rather than panicking and quitting, just go ahead and give it a name and variable and carry it though your calculation. You could report your final answer in terms of this variable or, better yet, the quantity may cancel out during the calculation. For example, we did not know the current in the solenoid, but needed to carry it through our calculation. Upon defining it to be $i_{solenoid}$ and carrying it through, the current term from the mutual inductance ended up cancelling with the current term from the magnetic flux.

20.83

SET UP

In the real world, solenoids are made with wire, which has some resistance associated with it. Let's say a cylindrical solenoid ($l_{solenoid} = 35.0 \times 10^{-2}$ m, $d_{solenoid} = 4.50 \times 10^{-2}$ m) is made of copper wire ($\rho_{copper} = 1.725 \times 10^{-8}$ $\Omega \cdot$m, $d_{wire} = 0.6438 \times 10^{-3}$ m) that is wound about a central axis. The adjacent loops touch each other but do not overlap, which means the total length of the solenoid is equal to the number of loops multiplied by the diameter of the wire. The resistance of the solenoid is equal to the resistance of the entire length of wire; the length of the wire is equal to the number of loops multiplied by the circumference of the solenoid.

The resistance of the copper wire is equal to $R = \dfrac{\rho_{copper} l_{wire}}{A_{wire}}$. The inductance of a solenoid is $L = \dfrac{\mu_0 N^2 A_{solenoid}}{l_{solenoid}}$.

SOLVE

Part a)

$$l_{solenoid} = N d_{wire}$$

$$N = \frac{l_{solenoid}}{d_{wire}} = \frac{35.0 \times 10^{-2} \text{ m}}{0.6438 \times 10^{-3} \text{ m}} = \boxed{544 \text{ loops}}$$

Part b)

Resistance:

$$R = \frac{\rho_{copper} l_{wire}}{A_{wire}} = \frac{\rho_{copper}(N(\pi d_{solenoid}))}{\left(\pi \left(\frac{d_{wire}}{2}\right)^2\right)} = \frac{4 N \rho_{copper} d_{solenoid}}{d_{wire}^2}$$

$$= \frac{4(544)(1.725 \times 10^{-8} \ \Omega \cdot \text{m})(4.50 \times 10^{-2} \text{ m})}{(0.6438 \times 10^{-3} \text{ m})^2} = \boxed{4.07 \ \Omega}$$

Inductance:

$$L = \frac{\mu_0 N^2 A_{solenoid}}{l_{solenoid}} = \frac{\mu_0 N^2 \left(\pi \left(\frac{d_{solenoid}}{2}\right)^2\right)}{l_{solenoid}} = \frac{\mu_0 \pi N^2 d_{solenoid}^2}{4 l_{solenoid}}$$

$$= \frac{\left(4\pi \times 10^{-7} \frac{\text{H}}{\text{m}}\right)\pi(544)^2(4.50 \times 10^{-2} \text{ m})^2}{4(35.0 \times 10^{-2} \text{ m})} = \boxed{1.69 \times 10^{-3} \text{ H} = 1.69 \text{ mH}}$$

REFLECT

Copper wire is known to be a good conductor so we would expect the resistance of the solenoid to be small.

Chapter 21
AC Circuits

Conceptual Questions

21.1 It means that the current peaks one-fourth of a period after the voltage drop peaks. One-fourth of a period corresponds to a phase difference of 90 degrees.

21.11 The electrical energy stored in the capacitor is $U_E = \frac{1}{2}\frac{1}{C}q^2$, which varies as $\cos^2(\omega t)$,

and the magnetic energy stored in the inductor is $U_L = \frac{1}{2}Li^2$, which varies as $\sin^2(\omega t)$.

Therefore, when a maximum amount of energy is stored in the capacitor, no energy is stored in the inductor, and *vice versa*.

21.15 An inductor with resistance is equivalent to a series LR combination. The tangent of the phase angle equals the ratio of the inductive resistance to the resistance. The phase angle varies with frequency because the inductive resistance does.

Multiple-Choice Questions

21.21 D ($2\sqrt{2}V$). The peak-to-peak voltage is twice the peak voltage. The peak voltage is $\sqrt{2}$ times the root-mean-square voltage.

21.27 A (the phase angle is zero).

$$\tan(\varphi) = \frac{X_L - X_C}{R} = 0$$

$$\varphi = \arctan(0) = 0$$

Estimation/Numerical Questions

21.31 Most household appliances draw between 1 and 20 amps. Most households have a main circuit breaker that "trips" at 150 A or so.

Problems

21.37

SET UP

The peak voltage across the terminals of a sinusoidal AC source ($f = 60$ Hz) is $V_0 = 17$ V. The voltage at $t = 0$ is equal to 0, which means we can model the voltage as a function of time by

$V(t) = V_0 \sin(\omega t) = V_0 \sin((2\pi f)t)$. Using the provided information and the algebraic form for $V(t)$, we can solve for the voltage at $t = 2.00 \times 10^{-3}$ s.

SOLVE

$$V(t = 2.00 \times 10^{-3} \text{ s}) = (17 \text{ V}) \sin(2\pi(60 \text{ Hz})(2.00 \times 10^{-3} \text{ s})) = \boxed{12 \text{ V}}$$

REFLECT

The maximum voltage occurs at one-quarter cycle or $t = 0.0042$ s, which means $V(t = 0.002 \text{ s})$ should be positive and less than 17 V.

21.45

SET UP

The primary coil ($N_p = 400$ turns) of a step-down transformer is connected to an AC line that has an applied voltage of $V_p = 120$ V. The secondary coil voltage is $V_s = 6.50$ V. The relationship between the number of turns and the voltages of each coil is $V_s = \dfrac{N_s}{N_p} V_p$. We can rearrange this equation to calculate the number of turns of the secondary coil N_s.

SOLVE

$$V_s = \frac{N_s}{N_p} V_p$$

$$N_s = \frac{V_s N_p}{V_p} = \frac{(6.50 \text{ V})(400 \text{ turns})}{120 \text{ V}} = 21.7 = \boxed{22 \text{ turns}}$$

REFLECT

We are told this is a step-down transformer, which means the $N_s < N_p$.

Get Help: P'Cast 21.2 – High Voltage Transformer

21.49

SET UP

The primary coil ($N_p = 400$ turns) of a step-down transformer is connected to an AC line that has an applied voltage of $V_p = 120$ V. The secondary coil voltage is $V_s = 6.30$ V. The relationship between the number of turns and the voltages of each coil is $V_s = \dfrac{N_s}{N_p} V_p$. We can rearrange this equation to calculate the number of turns of the secondary coil N_s. The secondary coil supplies a current of $i_s = 15.0$ A. We can use the relationship between the currents and the voltages of the two coils, $i_s = \dfrac{V_p}{V_s} i_p$, to calculate the current in the primary coil.

SOLVE

Part a)

$$V_s = \frac{N_s}{N_p}V_p$$

$$N_s = \frac{V_s N_p}{V_p} = \frac{(6.30 \text{ V})(400 \text{ turns})}{120 \text{ V}} = \boxed{21 \text{ turns}}$$

Part b)

$$i_s = \frac{V_p}{V_s}i_p$$

$$i_p = \frac{V_s}{V_p}i_s = \left(\frac{6.30 \text{ V}}{120 \text{ V}}\right)(15.0 \text{ A}) = \boxed{0.788 \text{ A}}$$

REFLECT

A step-down transformer is designed such that the primary coil has a high voltage but low current and the secondary coil has a low voltage but high current.

Get Help: P'Cast 21.2 – High Voltage Transformer

21.53

SET UP

At $t = 2.0 \times 10^{-3}$ s, a voltage phasor has a value of 50 V and makes an angle of $\frac{\pi}{4}$ rad with the x-axis. This angle is equal to ωt, where ω is the angular frequency of the voltage. The voltage at any given time $V(t)$ is equal to the projection of the phasor onto the y-axis, $V(t) = V_0 \sin(\omega t)$, where V_0 is the peak voltage.

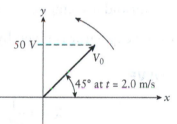

Figure 21-1 Problem 53

SOLVE

Angular frequency:

$$\omega t = \frac{\pi}{4}$$

$$\omega = \frac{\left(\frac{\pi}{4}\right)}{t} = \frac{\left(\frac{\pi}{4}\right)}{2.0 \times 10^{-3} \text{ s}} = \boxed{390 \frac{\text{rad}}{\text{s}}}$$

Peak voltage:

$$V(t) = V_0 \sin(\omega t)$$

$$V_0 = \frac{V(t)}{\sin(\omega t)} = \frac{50 \text{ V}}{\sin\left(\frac{\pi}{4}\right)} = \frac{50 \text{ V}}{\left(\frac{1}{\sqrt{2}}\right)} = \boxed{71 \text{ V}}$$

REFLECT
Remember to convert the angle into radians before performing a calculation.

21.57

SET UP
A circuit that contains a resistor, an inductor ($L = 5.00$ H), and a capacitor resonates at $f_0 = 1000$ Hz. Using the expression for the natural frequency of an LRC circuit, $\omega_0 = \sqrt{\dfrac{1}{LC}}$, we can calculate the value of the capacitor. Recall that $\omega_0 = 2\pi f_0$.

SOLVE

$$\omega_0 = \sqrt{\frac{1}{LC}}$$

$$C = \frac{1}{L\omega_0^2} = \frac{1}{L(2\pi f_0)^2} = \frac{1}{4\pi^2(5.00\ \text{H})(1000\ \text{Hz})^2} = \boxed{5.07 \times 10^{-9}\ \text{F} = 5.07\ \text{nF}}$$

REFLECT
The resistor does not affect the natural frequency of the circuit, only the maximum current.

21.61

SET UP
A sinusoidal voltage ($f = 60$ Hz) is applied to a capacitor ($C = 50.0 \times 10^{-6}$ F). The reactance of the capacitor is given by $X_C = \dfrac{1}{\omega C} = \dfrac{1}{(2\pi f)C}$.

SOLVE

$$X_C = \frac{1}{\omega C} = \frac{1}{(2\pi f)C} = \frac{1}{2\pi(60\ \text{Hz})(50.0 \times 10^{-6}\ \text{F})} = \boxed{53\ \Omega}$$

REFLECT
The reactance of a capacitor or inductor is equivalent to the resistance of a resistor, but the reactance depends upon the frequency of the applied voltage, while we usually assume the resistance remains constant.

Get Help: P'Cast 21.3 – High-Pass Filter

21.65

SET UP
Voltages oscillating at frequencies of $f = 60$ Hz, 6000 Hz, and 6×10^6 Hz are applied to an inductor ($L = 5.0 \times 10^{-3}$ H). The resulting reactance of the inductor is given by $X_L = \omega L = (2\pi f)L$.

SOLVE

Part a)

$$X_L = 2\pi(60\ \text{Hz})(5.0 \times 10^{-3}\ \text{H}) = \boxed{1.88\ \Omega}$$

Part b)

$$X_L = 2\pi(6000 \text{ Hz})(5.0 \times 10^{-3} \text{ H}) = \boxed{188 \ \Omega}$$

Part c)

$$X_L = 2\pi(6 \times 10^6 \text{ Hz})(5.0 \times 10^{-3} \text{ H}) = \boxed{1.88 \times 10^5 \ \Omega = 188 \text{ k}\Omega}$$

REFLECT

We could have also used proportional reasoning to calculate the answers to parts b and c. The frequency in part b is 100 times larger than the frequency in part a, which means the reactance will also be 100 times larger. The frequency in part c is 1000 times larger than the frequency in part b, which means the reactance will also be 1000 times larger.

Get Help: P'Cast 21.3 – High-Pass Filter

21.69

SET UP

A sinusoidal voltage ($V_{rms} = 40.0$ V, $f = 100$ Hz) is applied to a resistor ($R = 100 \ \Omega$), an inductor ($L = 0.200$ H), and a capacitor ($C = 50.0 \times 10^{-6}$ F). The current as a function of time for a resistor, inductor, and capacitor each attached separately to an AC voltage source is $i_R(t) = \dfrac{V_0}{R} \sin(\omega t)$, $i_L(t) = \dfrac{V_0}{\omega L} \sin\left(\omega t - \dfrac{\pi}{2}\right)$, and $i_C(t) = \omega C V_0 \sin\left(\omega t + \dfrac{\pi}{2}\right)$, respectively. The

peak current is the amplitude of the oscillation in all cases. The average power delivered to the resistor is equal to $P_{R, avg} = \dfrac{1}{2} \dfrac{V_0^2}{R}$. Since the power is equal to the voltage multiplied by the current, the average power delivered to the inductor and capacitor is zero because the current and voltage are 90 degrees out of phase.

SOLVE

Part a)

Peak current:

$$i_R(t) = \frac{V_0}{R} \sin(\omega t)$$

$$i_{R, \text{ max}} = \frac{V_0}{R} = \frac{V_{rms}\sqrt{2}}{R} = \frac{(40.0 \text{ V})\sqrt{2}}{100 \ \Omega} = \boxed{0.566 \text{ A}}$$

Average power:

$$P_{R, \text{ avg}} = \frac{1}{2} \frac{V_0^2}{R} = \frac{1}{2} \frac{(V_{rms}\sqrt{2})^2}{R} = \frac{V_{rms}^2}{R} = \frac{(40.0 \text{ V})^2}{100 \ \Omega} = \boxed{16.0 \text{ W}}$$

Part b)

Peak current:

$$i_L(t) = \frac{V_0}{\omega L} \sin\left(\omega t - \frac{\pi}{2}\right)$$

$$i_{L,\,max} = \frac{V_0}{\omega L} = \frac{V_{rms}\sqrt{2}}{(2\pi f)L} = \frac{(40.0\text{ V})\sqrt{2}}{2\pi(100\text{ Hz})(0.200\text{ H})} = \boxed{0.450\text{ A}}$$

Average power:

$$P_{C,\,avg} = 0$$

Part c)

Peak current:

$$i_C(t) = \omega C V_0 \sin\left(\omega t + \frac{\pi}{2}\right)$$

$$i_{C,\,max} = \omega C V_0 = (2\pi f)C(V_{rms}\sqrt{2}) = 2\pi\sqrt{2}fCV_{rms}$$

$$= 2\pi\sqrt{2}(100\text{ Hz})(50.0 \times 10^{-6}\text{ F})(40.0\text{ V}) = \boxed{1.78\text{ A}}$$

Average power:

$$P_{C,\,avg} = 0$$

REFLECT

The average of a function is $f_{avg} = \frac{1}{b-a}\int_a^b f(x)dx$, so the time average of the power delivered to a circuit element is $P_{avg} = \frac{1}{T}\int_0^T i(t)V(t)dt$. For a capacitor and an inductor, the integrand is proportional to $\sin(\omega t)\cos(\omega t)$, so the integral exactly equals zero.

Get Help: P'Cast 21.3 – High-Pass Filter

21.71

SET UP

An AC voltage, described by $V(t) = (10\text{ V})\sin(12\pi t)$, is applied to an *LRC* series circuit with $L = 0.250$ H, $R = 20\ \Omega$, and $C = 350 \times 10^{-6}$ F. We want to know the instantaneous voltage across each element at a time $t = 0.04$ s. First we need to find the peak current in the circuit from the peak voltage and the impedance. Next, the phase angle between the time-varying current and the time-varying voltage is given by $\tan(\varphi) = \dfrac{\omega L - \left(\frac{1}{\omega C}\right)}{R}$, where $\omega = 12\pi\dfrac{\text{rad}}{\text{s}}$.

Finally, the voltage across the resistor, capacitor, and inductor in the *LRC* circuit as a function of time are given by $V_R(t) = i_0 R \sin(12\pi t + |\varphi|)$, $V_C(t) = \dfrac{i_0}{\omega C}\sin\left(12\pi t + |\varphi| - \dfrac{\pi}{2}\right)$, and $V_L(t) = i_0\omega L \sin\left(12\pi t + |\varphi| + \dfrac{\pi}{2}\right)$, respectively.

SOLVE

Peak current:

$$i_0 = \frac{V_0}{Z} = \frac{V_0}{\sqrt{\left(\omega L - \dfrac{1}{\omega C}\right)^2 + R^2}}$$

$$= \frac{10 \text{ V}}{\sqrt{\left(\left(12\pi \dfrac{\text{rad}}{\text{s}}\right)(0.250 \text{ H}) - \left(\dfrac{1}{\left(12\pi \dfrac{\text{rad}}{\text{s}}\right)(350 \times 10^{-6} \text{ F})}\right)\right)^2 + (20 \ \Omega)^2}}$$

$$= \frac{10 \text{ V}}{69.3 \ \Omega} = 0.1443 \text{ A}$$

Phase angle:

$$\tan(\varphi) = \frac{\omega L - \left(\dfrac{1}{\omega C}\right)}{R}$$

$$\varphi = \arctan\left(\frac{\omega L - \left(\dfrac{1}{\omega C}\right)}{R}\right)$$

$$= \arctan\left(\frac{\left(12\pi \dfrac{\text{rad}}{\text{s}}\right)(0.250 \text{ H}) - \left(\dfrac{1}{\left(12\pi \dfrac{\text{rad}}{\text{s}}\right)(350 \times 10^{-6} \text{ F})}\right)}{20 \ \Omega}\right) = -1.278 \text{ rad}$$

Voltage across the resistor at $t = 0.04$ s:

$$V_R(t) = i_0 R \sin(12\pi t + |\varphi|) \text{ (SI units)}$$

$$V_R(0.04) = (0.1443)(20) \sin(12\pi(0.04) + 1.278) = \boxed{1.004 \text{ V}}$$

Voltage across the capacitor at $t = 0.04$ s:

$$V_C(t) = \frac{i_0}{\omega C} \sin\left(12\pi t + |\varphi| - \frac{\pi}{2}\right) \text{ (SI units)}$$

$$V_C(0.04) = \frac{(0.1443)}{(12\pi)(350 \times 10^{-6})} \sin\left(12\pi(0.04) + 1.278 - \frac{\pi}{2}\right) = \boxed{10.25 \text{ V}}$$

Voltage across the inductor at $t = 0.04$ s:

$$V_L(t) = i_0 \omega L \sin\left(12\pi t + |\varphi| + \frac{\pi}{2}\right) \text{ (SI units)}$$

$$V_L(0.04) = (0.1443)(12\pi)(0.250) \sin\left(12\pi(0.04) + 1.278 + \frac{\pi}{2}\right) = \boxed{-1.275 \text{ V}}$$

REFLECT

We can double check our answer to make sure the sum of the voltages across each element at $t = 0.04$ s equals the total voltage at $t = 0.04$ s:

$$V(t) = V_0 \sin(12\pi t)$$

$$V(0.04) = (10) \sin(12\pi(0.04)) = 9.98 \text{ V}$$

$$V_R(0.04) + V_C(0.04) + V_L(0.04) = (1.004 \text{ V}) + (10.25 \text{ V}) + (-1.275 \text{ V}) = 9.98 \text{ V}$$

21.75

SET UP

In an LRC series circuit, the inductive reactance is $X_L = 2500 \ \Omega$, the capacitive reactance is $X_C = 3400 \ \Omega$, and the resistance is $R = 1000 \ \Omega$. The impedance of the circuit is given by $Z = \sqrt{(X_L - X_C)^2 + R^2}$.

SOLVE

$$Z = \sqrt{(X_L - X_C)^2 + R^2} = \sqrt{((2500 \ \Omega) - (3400 \ \Omega))^2 + (1000 \ \Omega)^2} = \boxed{1345 \ \Omega}$$

REFLECT

Since $X_C > X_L$ the capacitor will have a greater effect on the current than the inductor, so this circuit is more "capacitive" than "inductive".

Get Help: P'Cast 21.4 – A Series LRC Circuit Driven by an AC Voltage – Impedance

21.79

SET UP

A resistor ($R = 250 \ \Omega$) and an inductor ($L = 0.04$ H) are wired in series with an AC voltage source ($V_{max} = 100$ V). The inductance of the circuit is given by $Z = \sqrt{(\omega L)^2 + R^2} = \sqrt{(2\pi f L)^2 + R^2}$, since the capacitance in the circuit is negligible. The peak current in the circuit is equal to the peak input voltage divided by the impedance. Once we have the current in terms of V_{max}, we can use the expression for the voltage across the inductor $V_{L, max} = i_{0, max} X_L$ to solve for the peak voltage across the inductor for $f = 50$ Hz and $f = 5000$ Hz. We can determine what this circuit filters by comparing the voltage across the inductor to the input voltage as a function of frequency and see if there is a pattern.

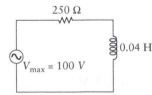

Figure 21-2 Problem 79

SOLVE

Inductance:

$$Z = \sqrt{(\omega L)^2 + R^2} = \sqrt{(2\pi f L)^2 + R^2}$$

Current:

$$i_{0,\,\text{max}} = \frac{V_{\text{in, max}}}{Z} = \frac{V_{\text{max}}}{\sqrt{(2\pi fL)^2 + R^2}}$$

Peak voltage across the inductor:

$$V_{\text{L, max}} = i_{0,\,\text{max}} X_{\text{L}} = \left(\frac{V_{\text{max}}}{\sqrt{(2\pi fL)^2 + R^2}}\right)(2\pi fL) = \frac{2\pi fL}{\sqrt{(2\pi fL)^2 + R^2}} V_{\text{max}}$$

Part a)

$$V_{\text{L, max at 50 Hz}} = \frac{2\pi(50 \text{ Hz})(0.04 \text{ H})}{\sqrt{(2\pi(50 \text{ Hz})(0.04 \text{ H}))^2 + (250 \text{ }\Omega)^2}}(100 \text{ V}) = \boxed{5.02 \text{ V} = 0.0502 V_{\text{max}}}$$

Part b)

$$V_{\text{L, max at 5000 Hz}} = \frac{2\pi(5000 \text{ Hz})(0.04 \text{ H})}{\sqrt{(2\pi(5000 \text{ Hz})(0.04 \text{ H}))^2 + (250 \text{ }\Omega)^2}}(100 \text{ V}) = \boxed{98.1 \text{ V} = 0.981 V_{\text{max}}}$$

Part c) Looking at our answers for parts (a) and (b), this *LR* circuit $\boxed{\text{attenuates low-frequency signals, while allowing high-frequency signals to pass}}$. (That is, the voltage across the inductor is much smaller than V_{in} for low-frequency signals and is on the order of V_{in} for high-frequency signals.) This circuit acts as a high pass filter.

REFLECT

A low pass filter allows low-frequency signals to pass and attenuates high-frequency signals.

21.83

SET UP

A circuit contains a resistor, an inductor ($L = 6.0$ H), and a capacitor ($C = 5.0 \times 10^{-9}$ F) wired in series. Using the expression for the natural frequency of an *LRC* circuit, $\omega_0 = \sqrt{\dfrac{1}{LC}}$, we can calculate the resonant frequency of the circuit. Recall that $\omega_0 = 2\pi f_0$.

SOLVE

$$\omega_0 = 2\pi f_0 = \sqrt{\frac{1}{LC}}$$

$$f_0 = \frac{1}{2\pi}\sqrt{\frac{1}{LC}} = \frac{1}{2\pi}\sqrt{\frac{1}{(6.0 \text{ H})(5.0 \times 10^{-9} \text{ F})}} = \boxed{920 \text{ Hz}}$$

REFLECT

The resistor does not affect the natural frequency of the circuit, only the maximum current.

Get Help: P'Cast 21.9 – Greatest Hits 99.7

21.85

SET UP

We are asked to determine an expression for the root-mean-square voltage associated with a specific plot of voltage vs. time (see figure). The signal repeats every T and varies linearly from $+V_0$ to $-V_0$, which means we can represent the voltage as a function of time by

$V(t) = \dfrac{-2V_0}{T}t + V_0$ for $0 < t < T$. The root-mean-square of the voltage involves first squaring the voltage, then finding its average, and finally taking the square root. As a reminder, the

average of a function over an interval from $t = a$ to $t = b$ is equal to $f_{avg} = \dfrac{1}{b-a}\displaystyle\int_a^b f(t)dt$.

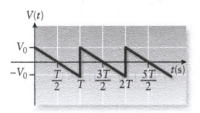

Figure 21-3 Problem 85

SOLVE

Root-mean-square:

$$V_{rms}^2 = \frac{1}{T}\int (V(t))^2 dt = \frac{1}{T}\int_0^T \left(\frac{-2V_0}{T}t + V_0\right)^2 dt = \frac{1}{T}\int_0^T \left(\frac{4V_0^2}{T^2}t^2 - \frac{4V_0^2}{T}t + V_0^2\right)dt$$

$$= \frac{V_0^2}{T}\int_0^T \left(\frac{4}{T^2}t^2 - \frac{4}{T}t + 1\right)dt = \frac{V_0^2}{T}\left[\frac{4}{3T^2}t^3 - \frac{2}{T}t^2 + t\right]_0^T = \frac{V_0^2}{T}\left[\frac{4}{3T^2}T^3 - \frac{2}{T}T^2 + T\right]$$

$$= \frac{V_0^2}{T}\left[\frac{4}{3}T - 2T + T\right] = \frac{V_0^2}{T}\left[\frac{1}{3}T\right] = \frac{V_0^2}{3}$$

$$V_{rms} = \sqrt{\frac{V_0^2}{3}} = \boxed{\frac{V_0}{\sqrt{3}}}$$

REFLECT

From dimensional analysis, there is no way the root-mean-square voltage can depend on the period.

21.91

SET UP

An AC generator is connected across a light bulb ($R = 8.50\ \Omega$). The electrical generator is made by rotating a flat coil in a uniform magnetic field ($B = 0.225$ T). The flat coil has 33

windings and is square, measuring 0.150 m on each side. The coil rotates at $\omega = 745\dfrac{\text{rev}}{\text{min}}$

about an axis perpendicular to the magnetic field and parallel to its two opposite sides. This orientation means the area vector will rotate around at a frequency of ω. We'll assume the magnetic field vector and area vector begin parallel. The alternating induced potential in the coil can be determined through Faraday's law. The amplitude of the resulting function is also equal to the voltage amplitude V_0 across the light bulb. The current amplitude for the bulb is equal to $i_0 = \dfrac{V_0}{R}$ through Ohm's law. The average rate heat is generated is equal to the average power emitted by the light bulb, $P_{avg} = i_{rms}^2 R$, where $i_{rms} = \dfrac{i_0}{\sqrt{2}}$. The energy consumed by the light bulb in an hour is equal to its power multiplied by 1 hour.

SOLVE

Part a)

Angular frequency:

$$\omega = 745\frac{\text{rev}}{\text{min}} \times \frac{1\ \text{min}}{60\ \text{s}} \times \frac{2\pi\ \text{rad}}{1\ \text{rev}} = \frac{745\pi}{30}\frac{\text{rad}}{\text{s}}$$

Induced potential:

$$V = -\frac{d\Phi_B}{dt} = -\frac{d}{dt}[NBA\ \cos(\omega t)] = -NBA\frac{d}{dt}[\cos(\omega t)] = NBA\omega\ \sin(\omega t)$$

$$= (33)(0.225\ \text{T})(0.150\ \text{m})^2\left(\frac{745\pi}{30}\frac{\text{rad}}{\text{s}}\right)\sin(\omega t) = (13.0\ \text{V})\sin(\omega t)$$

$$\boxed{V_0 = 13.0\ \text{V}}$$

Current:

$$i_0 = \frac{V_0}{R} = \frac{13.0\ \text{V}}{8.50\ \Omega} = \boxed{1.53\ \text{A}}$$

Part b)

$$P_{avg} = i_{rms}^2 R = \left(\frac{i_0}{\sqrt{2}}\right)^2 R = \frac{(1.53\ \text{A})^2}{2}(8.50\ \Omega) = \boxed{9.99\ \text{W}}$$

Part c)

$$E_{consumed} = Pt = (9.99\ \text{W})\left(1\ \text{hr} \times \frac{3600\ \text{s}}{1\ \text{hr}}\right) = \boxed{3.60 \times 10^4\ \text{J}}$$

REFLECT

In the United States, we buy light bulbs based on the power they consume at a voltage of 120 V rather than their resistance. Accordingly, a 60-W light bulb has a resistance of 240 Ω.

21.95

SET UP

An AC voltage source is connected to a resistor ($R = 5\ \Omega$), a capacitor ($C = 400 \times 10^{-6}$ F), and an inductor ($L = 0.025$ H), all wired in series. The current in the circuit is $i(t) = (10\ \text{A})\sin(120\pi t)$, which means $\omega = 120\pi\dfrac{\text{rad}}{\text{s}}$. The total impedance of this circuit can be calculated from $Z = \sqrt{\left(\omega L - \dfrac{1}{\omega C}\right)^2 + R^2}$. The current and voltage are in phase for

a resistor, which means $\varphi_R = 0$. For a capacitor, the voltage lags the current by $\dfrac{\pi}{2}$ rad, so

$\varphi_C = -\dfrac{\pi}{2}$. Finally, the voltage leads the current by $\dfrac{\pi}{2}$ rad in an inductor, which means $\varphi_L = \dfrac{\pi}{2}$.

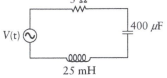

Figure 21-4 Problem 95

SOLVE

Part a)

$$Z = \sqrt{\left(\omega L - \frac{1}{\omega C}\right)^2 + R^2}$$

$$= \sqrt{\left(\left(120\pi\frac{\text{rad}}{\text{s}}\right)(0.025\ \text{H}) - \left(\frac{1}{\left(120\pi\frac{\text{rad}}{\text{s}}\right)(400 \times 10^{-6}\ \text{F})}\right)\right)^2 + (5\ \Omega)^2} = \boxed{5.73\ \Omega}$$

Part b)

$$\varphi_R = 0$$

$$\varphi_C = -\frac{\pi}{2}$$

$$\varphi_L = \frac{\pi}{2}$$

REFLECT

The current through all three circuit elements is equal since they are wired up in series. This is an inductive circuit since $X_L > X_C$.

Get Help: Interactive Example – AC Circuit 1

P'Cast 21.4 – A Series LRC Circuit Driven by an AC

Voltage – Impedance

21.99

SET UP

The current in a circuit consisting of an AC power supply and a capacitor is found to be $i(t) = i_0 \sin(\omega t)$. The current in the circuit is equal to the first derivative of the charge on the capacitor with respect to time. Assuming the charge is equal to zero at $t = 0$, we can solve for the charge as a function of time via separation of variables.

SOLVE

$$i(t) = \frac{dq}{dt} = i_0 \sin(\omega t)$$

$$\int_{q(0)}^{q(t)} dq = \int_0^t i_0 \sin(\omega t) dt$$

$$q(t) - q(0) = i_0 \left[-\frac{1}{\omega} \cos(\omega t) \right]_0^t$$

$$q(t) - 0 = -\frac{i_0}{\omega} [\cos(\omega t) - \cos(0)]_0^t$$

$$q(t) = -\frac{i_0}{\omega} [\cos(\omega t) - 1] = \boxed{\frac{i_0}{\omega} [1 - \cos(\omega t)]}$$

REFLECT

The charge on the capacitor oscillates in time between a minimum value of 0 and a maximum value of $\frac{2i_0}{\omega}$.

Get Help: P'Cast 21.3 – High-Pass Filter

21.103

SET UP

The circuit seen in the figure is known as a low pass filter. The input is an AC signal composed of many frequencies. The output detected is the voltage across the capacitor. The current in the circuit is equal to the input voltage divided by the impedance of the circuit. Once we have the current in terms of V_i, we can use the expression for the voltage across the capacitor $V_o = i_0 X_C$ to solve for the ratio between the output and input voltages.

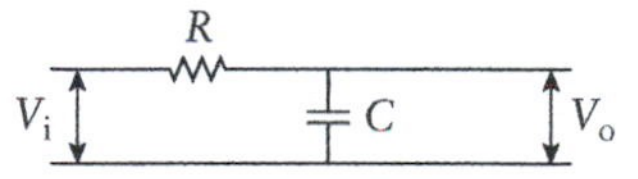

Figure 21-5 Problem 103

SOLVE

Current:

$$i_0 = \frac{V_i}{Z} = \frac{V_i}{\sqrt{\left(\frac{1}{\omega C}\right)^2 + R^2}}$$

Voltage across the capacitor:

$$V_o = i_0 X_C = \left(\frac{V_i}{\sqrt{\left(\frac{1}{\omega C}\right)^2 + R^2}} \right) \left(\frac{1}{\omega C} \right)$$

Ratio between V_o and V_i:

$$\frac{V_o}{V_i} = \frac{1}{\omega C \sqrt{\left(\frac{1}{\omega C}\right)^2 + R^2}} = \frac{1}{\sqrt{(\omega C)^2 \left(\frac{1}{\omega C}\right)^2 + (\omega C)^2 (R^2)}} = \boxed{\frac{1}{\sqrt{1 + (2\pi f R C)^2}}}$$

REFLECT

Looking at our final expression for the ratio of the voltages, we see the $\frac{V_o}{V_i}$ gets small as f gets large, which means the amplitude of the high-frequency components of the input signal is very small. This is consistent with our expectation that a low pass filter allows low-frequency signals to pass (relatively) unaffected while attenuating the high-frequency ones.

Get Help: P'Cast 21.9 – Greatest Hits 99.7

Chapter 22
Electromagnetic Waves

Conceptual Questions

22.3 The electric field is perpendicular to the magnetic field and the velocity of the wave. That means that the electric fields will vibrate, say, along the x-axis, the magnetic field along the y-axis, and the EM radiation will move along the z-axis. These three vectors are mutually perpendicular.

22.7 All EM waves move at the same speed ($c = 3.00 \times 10^8$ m/s). However, the wavelength times the frequency equals this constant ($c = \lambda f$). Therefore, the wavelength can vary (smaller/larger) in proportion to the frequency (larger/smaller).

Multiple-Choice Questions

22.13 C (have same speed). All EM radiation travels at the speed of light.

Get Help: P'Cast 22.1 – Visible Light

22.15 D (sound). EM waves can travel in a vacuum.

22.19 E (to both time-independent and time-dependent electric and magnetic fields). Maxwell's equations are the fundamental relationships underlying all electric, magnetic, and electromagnetic phenomena, regardless of their time dependence.

Estimation/Numerical Questions

22.23 Green light has a wavelength around 550 nm. The energy of a green photon is then

$$E = \frac{hc}{\lambda} = \frac{(6.63 \times 10^{-34} \text{ J} \cdot \text{s})\left(3.00 \times 10^8 \frac{\text{m}}{\text{s}}\right)}{550 \times 10^{-9} \text{ m}} = 3.62 \times 10^{-19} \text{ J}$$

Get Help: P'Cast 22.1 – Visible Light

22.27 The people 300 km away receive the news first:

$$t_{\text{radio}} = \frac{\Delta x}{c} = \frac{300 \times 10^3 \text{ m}}{\left(3.00 \times 10^8 \frac{\text{m}}{\text{s}}\right)} = 1 \times 10^{-3} \text{ s}$$

$$t_{\text{sound}} = \frac{\Delta x}{v} = \frac{3 \text{ m}}{\left(340 \frac{\text{m}}{\text{s}}\right)} = 8.8 \times 10^{-3} \text{ s}$$

Problems

22.31

SET UP

We are given a list of the wavelengths of some photons and asked to calculate their frequencies. The frequency of a photon is related to the wavelength by the speed of light,

$$f = \frac{c}{\lambda}.$$

SOLVE

A) $f = \dfrac{c}{\lambda} = \dfrac{\left(3.00 \times 10^8 \frac{m}{s}\right)}{700 \times 10^{-9} \ m} = \boxed{4.29 \times 10^{14} \ Hz}$

B) $f = \dfrac{c}{\lambda} = \dfrac{\left(3.00 \times 10^8 \frac{m}{s}\right)}{600 \times 10^{-9} \ m} = \boxed{5.00 \times 10^{14} \ Hz}$

C) $f = \dfrac{c}{\lambda} = \dfrac{\left(3.00 \times 10^8 \frac{m}{s}\right)}{500 \times 10^{-9} \ m} = \boxed{6.00 \times 10^{14} \ Hz}$

D) $f = \dfrac{c}{\lambda} = \dfrac{\left(3.00 \times 10^8 \frac{m}{s}\right)}{400 \times 10^{-9} \ m} = \boxed{7.50 \times 10^{14} \ Hz}$

E) $f = \dfrac{c}{\lambda} = \dfrac{\left(3.00 \times 10^8 \frac{m}{s}\right)}{100 \times 10^{-9} \ m} = \boxed{3.00 \times 10^{15} \ Hz}$

F) $f = \dfrac{c}{\lambda} = \dfrac{\left(3.00 \times 10^8 \frac{m}{s}\right)}{0.033 \times 10^{-9} \ m} = \boxed{9.01 \times 10^{18} \ Hz}$

G) $f = \dfrac{c}{\lambda} = \dfrac{\left(3.00 \times 10^8 \frac{m}{s}\right)}{500 \times 10^{-6} \ m} = \boxed{6.00 \times 10^{11} \ Hz}$

H) $f = \dfrac{c}{\lambda} = \dfrac{\left(3.00 \times 10^8 \frac{m}{s}\right)}{63.3 \times 10^{-12} \ m} = \boxed{4.74 \times 10^{18} \ Hz}$

REFLECT

The frequency and wavelength are inversely proportional to one another, so a smaller wavelength will have a larger frequency, and *vice versa*.

Get Help: P'Cast 22.1 – Visible Light

22.37

SET UP

The speed of light in a vacuum is $c = 3.00 \times 10^8$ m/s. The distance light travels in 10×10^{-9} s can be found by multiplying the speed by the time interval.

SOLVE

$$c = \frac{\Delta x}{\Delta t}$$

$$\Delta x = c(\Delta t) = \left(3.00 \times 10^8 \frac{m}{s}\right)(10 \times 10^{-9} \text{ s}) = \boxed{3 \text{ m}}$$

REFLECT

One light-year is the distance light travels during one year: $1 \text{ ly} = 9.4607 \times 10^{12}$ km.

22.41

SET UP

We can rewrite the speed of light in terms of the angular frequency ω and the wavenumber k, $c = \frac{\omega}{k}$, in order to solve for the wavenumber of a photon with $\omega = 6.28 \times 10^{15}\frac{rad}{s}$ or the angular frequency of a photon with $k = 4\pi \times 10^6\frac{rad}{m}$. The frequency is equal to $f = \frac{\omega}{2\pi}$, and the wavenumber is related to the wavelength though its definition, $k = \frac{2\pi}{\lambda}$.

SOLVE

Part a)

$$c = \frac{\omega}{k}$$

$$k = \frac{\omega}{c} = \frac{\left(6.28 \times 10^{15}\frac{rad}{s}\right)}{\left(3.00 \times 10^8\frac{m}{s}\right)} = \boxed{2.09 \times 10^7\frac{rad}{m}}$$

Part b)

Angular frequency:

$$c = \frac{\omega}{k}$$

$$\omega = kc = \left(3.00 \times 10^8\frac{m}{s}\right)\left(4\pi \times 10^6\frac{rad}{m}\right) = \boxed{3.77 \times 10^{15}\frac{rad}{s}}$$

Frequency:

$$f = \frac{\omega}{2\pi} = \frac{\left(3.77 \times 10^{15}\frac{rad}{s}\right)}{2\pi} = \boxed{6.00 \times 10^{14} \text{ Hz}}$$

Wavelength:

$$k = \frac{2\pi}{\lambda}$$

$$\lambda = \frac{2\pi}{k} = \frac{2\pi}{\left(4\pi \times 10^6 \frac{\text{rad}}{\text{m}}\right)} = \boxed{5.00 \times 10^{-7} \text{ m}}$$

REFLECT
The photon in part (a) is 300-nm ultraviolet light. The photon in part (b) is blue-green visible light.

22.45

SET UP
Charge is flowing onto the positive plate and off of the negative plate of a parallel plate capacitor with closely spaced plates at a rate of $\frac{dq}{dt} = 2.8$ A. If we treat the parallel plates as being approximately infinite, the electric field in between the plates has a magnitude of $E = \frac{q}{A\varepsilon_0}$, where A is the cross-sectional area. The displacement current through the capacitor between the plates is equal to $\varepsilon_0 \frac{d\Phi_E}{dt}$, where $\Phi_E = \vec{E} \cdot \vec{A}$.

SOLVE

$$\varepsilon_0 \frac{d\Phi_E}{dt} = \varepsilon_0 \frac{d}{dt}[\vec{E} \cdot \vec{A}] = \varepsilon_0 \frac{d}{dt}[EA] = \varepsilon_0 \frac{d}{dt}\left[\left(\frac{q}{A\varepsilon_0}\right)A\right] = \frac{dq}{dt} = \boxed{2.8 \text{ A}}$$

REFLECT
It makes sense that the displacement current "through" the capacitor should equal the rate of charge flowing onto the positive plate and off of the negative plate.

22.49

SET UP
We are asked to determine $\frac{\partial^2 f}{\partial t^2}$, where $f(x,t) = Ae^{\alpha t}\sin(kx - \omega t)$. In order to take the partial derivatives with respect to time, we need to hold the position fixed and treat it as a constant.

SOLVE

$$f(x,t) = Ae^{\alpha t}\sin(kx - \omega t)$$

$$\frac{\partial f}{\partial t} = \frac{\partial}{\partial t}[Ae^{\alpha t}\sin(kx - \omega t)] = A[\alpha e^{\alpha t}\sin(kx - \omega t) - e^{\alpha t}\omega\cos(kx - \omega t)]$$

$$= Ae^{\alpha t}[\alpha\sin(kx - \omega t) - \omega\cos(kx - \omega t)]$$

$$\frac{\partial^2 f}{\partial t^2} = \frac{\partial}{\partial t}[Ae^{\alpha t}[\alpha\sin(kx - \omega t) - \omega\cos(kx - \omega t)]]$$

$$= A\alpha e^{\alpha t}[\alpha\sin(kx - \omega t) - \omega\cos(kx - \omega t)] + Ae^{\alpha t}[-\alpha\omega\cos(kx - \omega t) - \omega^2\sin(kx - \omega t)]$$

$$\boxed{= \alpha^2 Ae^{\alpha t}\sin(kx - \omega t) - 2\alpha\omega Ae^{\alpha t}\cos(kx - \omega t) - \omega^2 Ae^{\alpha t}\sin(kx - \omega t)}$$

REFLECT

It's always easiest to first explicitly calculate the first derivative when trying to find the second derivative, rather than trying to do it all in one step.

22.53

SET UP

We are asked to determine whether the function $E(x,t) = E_0[\sin(kx - \omega t) + \cos(kx - \omega t)]$ satisfies the one-dimensional wave equation, $\dfrac{\partial^2 E}{\partial x^2} = \mu_0\varepsilon_0\dfrac{\partial^2 E}{\partial t^2}$, where $\omega = kc$ and $\mu_0\varepsilon_0 = \dfrac{1}{c^2}$. In order to take the partial derivatives with respect to position or time, we need to hold the time or position fixed, respectively, and treat it as a constant.

SOLVE

Taking derivatives:

$$E(x,t) = E_0[\sin(kx - \omega t) + \cos(kx - \omega t)]$$

$$\frac{\partial E}{\partial x} = \frac{\partial}{\partial x}[E_0[\sin(kx - \omega t) + \cos(kx - \omega t)]] = E_0[k\cos(kx - \omega t) - k\sin(kx - \omega t)]$$

$$\frac{\partial^2 E}{\partial x^2} = \frac{\partial}{\partial x}[E_0[k\cos(kx - \omega t) - k\sin(kx - \omega t)]] = E_0[-k^2\sin(kx - \omega t) - k^2\cos(kx - \omega t)]$$

$$= -k^2 E_0[\sin(kx - \omega t) + \cos(kx - \omega t)]$$

$$\frac{\partial E}{\partial t} = \frac{\partial}{\partial t}[E_0[\sin(kx - \omega t) + \cos(kx - \omega t)]] = E_0[-\omega\cos(kx - \omega t) + \omega\sin(kx - \omega t)]$$

$$\frac{\partial^2 E}{\partial t^2} = \frac{\partial}{\partial t}[E_0[-\omega\cos(kx - \omega t) + \omega\sin(kx - \omega t)]] = E_0[-\omega^2\sin(kx - \omega t) - \omega^2\cos(kx - \omega t)]$$

$$= -\omega^2 E_0[\sin(kx - \omega t) + \cos(kx - \omega t)]$$

Wave equation:

$$\frac{\partial^2 E}{\partial x^2} = \mu_0\varepsilon_0\frac{\partial^2 E}{\partial t^2}$$

$$-k^2 E_0[\sin(kx - \omega t) + \cos(kx - \omega t)] \overset{?}{=} \mu_0\varepsilon_0[-\omega^2 E_0[\sin(kx - \omega t) + \cos(kx - \omega t)]]$$

$$k^2 \overset{?}{=} \mu_0\varepsilon_0\omega^2$$

$$k^2 \overset{?}{=} \left(\frac{1}{c^2}\right)(kc)^2 = k^2$$

REFLECT

The functions $E(x,t) = E_0\sin(kx - \omega t)$, $E(x,t) = E_0\cos(kx - \omega t)$, or their linear combination, $E(x,t) = E_0[\sin(kx - \omega t) + \cos(kx - \omega t)]$, are all solutions to the one-dimensional wave equation.

22.57

SET UP

A laser is made up of a cylindrical beam of diameter $d = 0.750 \times 10^{-2}$ m. The energy is pulsed, lasting $\Delta t = 1.50 \times 10^{-9}$ s. Each pulse contains an energy of $E = 2.00$ J. The length (in meters) of the pulse can be found by multiplying the pulse duration by the speed of light c. Using this length, we can calculate the volume of the cylindrical pulse and then the energy per unit volume for each pulse.

SOLVE

Part a)

$$c = \frac{\Delta x}{\Delta t}$$

$$\Delta x = c\Delta t = \left(3.00 \times 10^8\frac{\text{m}}{\text{s}}\right)(1.50 \times 10^{-9}\text{ s}) = \boxed{0.45 \text{ m}}$$

Part b)

$$\frac{E}{V} = \frac{E}{\pi R^2(\Delta x)} = \frac{E}{\pi\left(\dfrac{d}{2}\right)^2(\Delta x)} = \frac{4E}{\pi d^2(\Delta x)} = \frac{4(2.00 \text{ J})}{\pi(0.750 \times 10^{-2}\text{ m})^2(0.45 \text{ m})} = \boxed{1.01 \times 10^5\frac{\text{J}}{\text{m}^3}}$$

REFLECT

As a rule of thumb, light travels a distance of 1 m in 3.3 ns.

Chapter 23
Wave Properties of Light

Conceptual Questions

23.3 The light coming from the Sun is refracted as it passes through Earth's atmosphere. Because of the wavelength dependence of the index of refraction, red light bends less than orange than yellow than green than blue than violet. In addition, bluer light is scattered more than red light as it passes through the atmosphere. Because of these effects, more red/orange colored light can pass into the shadow of Earth, making the Moon appear red.

23.9 When light enters a material with a negative index of refraction, the light refracts "back" away from the normal to the surface as seen in the picture below:

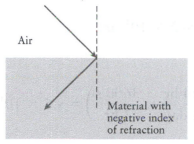

Air

Material with negative index of refraction

Figure 23-1 Problem 9

Get Help: P'Cast 23.1 – Seeing Under Water

23.15 Sunlight includes all of the colors of the rainbow and one of the delightful aspects of a diamond is its strong dispersion. The diamond bends violet light much more than red light so that sunlight is separated into its different colors as it passes through the diamond surface.

Multiple-Choice Questions

23.21 B $\left(\frac{1}{2}I_0\right)$. The intensity of unpolarized light drops by a factor of two when it passes through a linear polarizer. The second polarizer does not affect the intensity of the light because the two polarizers are aligned (*i.e.*, $\theta = 0$).

Get Help: Interactive Example – Polarization I
Interactive Example – Polarization II

23.25 A $(w_a > w_b)$. The width of the slit is inversely proportional to the fringe width.

(a)   (b)

Figure 23-2 Problem 25

Estimation/Numerical Questions

23.29 We can use $c = \dfrac{\Delta x}{\Delta t}$ to calculate the distance light travels.

Part a)

$$\Delta x = c(\Delta t) = \left(3.0 \times 10^8 \frac{m}{s}\right)(1 \text{ s}) = 3.0 \times 10^8 \text{ m}$$

Part b)

$$\Delta x = c(\Delta t) = \left(3.0 \times 10^8 \frac{m}{s}\right)(60 \text{ s}) = 2 \times 10^{10} \text{ m}$$

Part c)

$$\Delta x = c(\Delta t) = \left(3.0 \times 10^8 \frac{m}{s}\right)\left(1 \text{ yr} \times \frac{365.25 \text{ days}}{1 \text{ yr}} \times \frac{24 \text{ hr}}{1 \text{ day}} \times \frac{3600 \text{ s}}{1 \text{ hr}}\right) = 9.5 \times 10^{15} \text{ m}$$

Problems

23.35

SET UP

The speed of light in methylene iodide is $v_m = 1.72 \times 10^8 \frac{m}{s}$. The index of refraction of water is $n_w = 1.33$, which means the speed of light in water is $v_w = \dfrac{c}{n_w}$. We can use these data along with the definition of speed, $v = \dfrac{\Delta x}{\Delta t}$, to calculate the distance of methylene iodide that light must travel through such that it takes the same amount of time as light traveling through 1000×10^3 m of water.

SOLVE

$$v = \frac{\Delta x}{\Delta t}$$

$$\Delta t = \frac{\Delta x}{v}$$

$$\frac{(\Delta x)_m}{v_m} = \frac{(\Delta x)_w}{v_w}$$

$$(\Delta x)_m = \frac{(\Delta x)_w v_m}{v_w} = \frac{(\Delta x)_w v_m}{\left(\frac{c}{n_w}\right)} = \frac{(\Delta x)_w v_m n_w}{c}$$

$$= \frac{(1000 \times 10^3\,\text{m})\left(1.72 \times 10^8 \frac{\text{m}}{\text{s}}\right)(1.33)}{\left(3.0 \times 10^8 \frac{\text{m}}{\text{s}}\right)} = \boxed{7.63 \times 10^5\,\text{m} = 763\,\text{km}}$$

REFLECT

The index of refraction of methylene iodide is $n_m = \frac{c}{v_m} = \frac{\left(3.0 \times 10^8 \frac{\text{m}}{\text{s}}\right)}{\left(1.72 \times 10^8 \frac{\text{m}}{\text{s}}\right)} = 1.74$, which
means light will travel more slowly in methylene iodide compared to water. Therefore, we
expect the distance of methylene iodide should be smaller than the distance of water.

Get Help: Interactive Example – Refraction
P'Cast 23.1 – Seeing Under Water

23.39

SET UP

Total internal reflection occurs when the incident angle of light traveling from a medium with
index of refraction n_1 into a medium with index of refraction n_2 is greater than the critical
angle $\theta_c = \arcsin\left(\frac{n_2}{n_1}\right)$. It can only occur when $n_1 > n_2$. We can use the provided indices of
refraction to calculate the critical angle in each case.

SOLVE

A) $\theta_c = \arcsin\left(\frac{1.00}{1.50}\right) = \boxed{41.8°}$

B) $\theta_c = \arcsin\left(\frac{1.00}{1.33}\right) = \boxed{48.8°}$

C) $\theta_c = \arcsin\left(\frac{1.33}{1.56}\right) = \boxed{58.5°}$

D) $\boxed{\text{no critical angle}}$

REFLECT

Because $-1 \le \sin(\theta) \le 1$, $\arcsin\left(\frac{1.55}{1.00}\right) = \arcsin(1.55)$ is undefined.

Get Help: Picture It – Reflection and Refraction
P'Cast 23.2 – Critical Angle for a Glass-Air Boundary
P'Cast 23.3 – Critical Angles

23.43

SET UP

A scuba diver wants to look up from inside the water ($n_{\text{water}} = 1.33$) to see her friend standing on a very distant shore ($n_{\text{air}} = 1.00$). Since her friend is located so far away, the angle of the light ray is essentially 90 degrees. We can use Snell's law to calculate the angle the scuba diver must look.

SOLVE

$$n_1 \sin(\theta_1) = n_2 \sin(\theta_2)$$

$$(1.33) \sin(\theta_1) = (1.00) \sin(90°)$$

$$\theta_1 = \arcsin\left(\frac{1.00}{1.33}\right) = \boxed{48.8°}$$

REFLECT

This is equivalent to finding the critical angle for the system.

> **Get Help:** Picture It – Reflection and Refraction
> P'Cast 23.2 – Critical Angle for a Glass-Air Boundary
> P'Cast 23.3 – Critical Angles

23.49

SET UP

Blue and yellow light are incident on a glass slab ($t = 12$ cm) at an angle of 25 degrees relative to the normal. The index of refraction for the blue light in the glass is $n_{\text{glass, b}} = 1.545$, and the index of refraction for the yellow light in the glass is $n_{\text{glass, y}} = 1.523$. We are interested in the distance separating the two rays when they emerge from the other side of the slab. In order to calculate this, we first need to find the vertical displacement of each ray (either y_b or y_y for blue or yellow light, respectively) once it exits the slab. Applying Snell's law at the first interface will give the angle of the refracted light (either θ_b or θ_y for blue or yellow light, respectively) in the glass slab. The ray will travel at this angle until it reaches the other side. We can draw a triangle and relate the thickness of the slab t and the tangent of θ_b (or θ_y) to the vertical displacement y_b (or y_y). The distance between the rays is equal to the difference between the two vertical displacements.

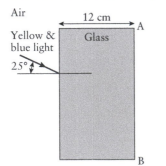

Figure 23-3 Problem 49

SOLVE
Blue light, angle of refraction:

$$n_{air} \sin(\theta_1) = n_{glass,\ b} \sin(\theta_b)$$

$$\theta_b = \arcsin\left(\frac{n_{air}}{n_{glass,\ b}} \sin(\theta_1)\right) = \arcsin\left(\left(\frac{1.000}{1.545}\right)\sin(25°)\right) = 15.87°$$

Blue light, vertical displacement:

Figure 23-4 Problem 49

$$\tan(\theta_b) = \frac{y_b}{t}$$

$$y_b = t\tan(\theta_b) = (12\ \text{cm})\tan(15.87°) = 3.4126\ \text{cm}$$

Yellow light, angle of refraction:

$$n_{air} \sin(\theta_1) = n_{glass,\ y} \sin(\theta_y)$$

$$\theta_y = \arcsin\left(\frac{n_{air}}{n_{glass,\ y}} \sin(\theta_1)\right) = \arcsin\left(\left(\frac{1.000}{1.523}\right)\sin(25°)\right) = 16.11°$$

Yellow light, vertical displacement:

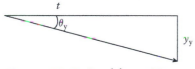

Figure 23-5 Problem 49

$$\tan(\theta_y) = \frac{y_y}{t}$$

$$y_y = t\tan(\theta_y) = (12\ \text{cm})\tan(16.11°) = 3.4660\ \text{cm}$$

Vertical difference between the blue and yellow rays:

$$\Delta y = y_y - y_b = (3.4660\ \text{cm}) - (3.4126\ \text{cm}) = \boxed{0.0534\ \text{cm}}$$

REFLECT
The thicker the piece of glass, the farther apart the blue and yellow rays will emerge from one another, which makes sense. The shallower the angle of the incoming light is, the closer together the rays will emerge, which also makes sense.

23.53

SET UP

Vertically polarized light with an initial intensity of I_0 is sent through a polarizing filter that makes a relative angle of θ to the filter. The intensity of the light after the polarizer is reduced by 25%, *i.e.*, $I = 0.75I_0$. The intensity of polarized light after it passes through a linear polarizer that makes a relative angle of θ is equal to $I = I_0\cos^2(\theta)$.

SOLVE

$$I = 0.75I_0 = I_0\cos^2(\theta)$$

$$\theta = \arccos(\pm\sqrt{0.75}) = \boxed{30°, \ 150°}$$

REFLECT

It makes sense that there should be two angles due to the symmetry of the system.

Get Help: Interactive Example – Polarization I
Interactive Example – Polarization II

23.59

SET UP

A person observes that the light rays from the Sun that bounce off the air-water surface are linearly polarized along the horizontal. When incoming light strikes the air-water boundary at Brewster's angle, $\theta_B = \arctan\left(\dfrac{n_{water}}{n_{air}}\right)$, the reflected light is completely polarized parallel to the surface. Brewster's angle is measured relative to the vertical, so the angle relative to the horizontal is equal to $\theta = 90° - \theta_B$.

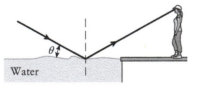

Water

Figure 23-6 Problem 59

SOLVE

Brewster's angle:

$$\theta_B = \arctan\left(\frac{n_{water}}{n_{air}}\right) = \arctan\left(\frac{1.33}{1.00}\right) = 53.1°$$

Angle above the horizontal:

$$\theta = 90° - \theta_B = 90° - 53.1° = \boxed{36.9°}$$

REFLECT

If the Sun is close to, but not exactly at, this angle, the reflected light will be strongly, not completely, polarized.

Get Help: P'Cast 23.6 – Brewster's Angle for Air to Water

23.63

SET UP

White light illuminates a thin film ($n_{film} = 1.35$) normal to the surface, and we observe that both blue light ($\lambda_b = 500$ nm) and red light ($\lambda_r = 700$ nm) are strongly reflected. The film has a thickness t and is floating on top of water ($n_{water} = 1.33$). The condition on the path difference for constructive interference in the case where the film has a higher index of refraction than the medium under it is $\left(m + \dfrac{1}{2} \right) \dfrac{\lambda_0}{n_{film}}$. Note that the wavelength of light changes when the index of refraction changes. We are interested in the wavelength of light in the film, so we need to divide the given wavelength by the index of refraction of the film. Since the thickness of the film is t, this path difference should be equal to $2t$. This same path difference strongly reflects *two* wavelengths of light, so we will have two expressions with different integers, m_b for the blue light and m_r for the red light. Setting the thicknesses equal to one another, we can solve the smallest integers that satisfy the resulting relationship. The minimum thickness of the film can be found by plugging the integer back into the constructive interference relationship.

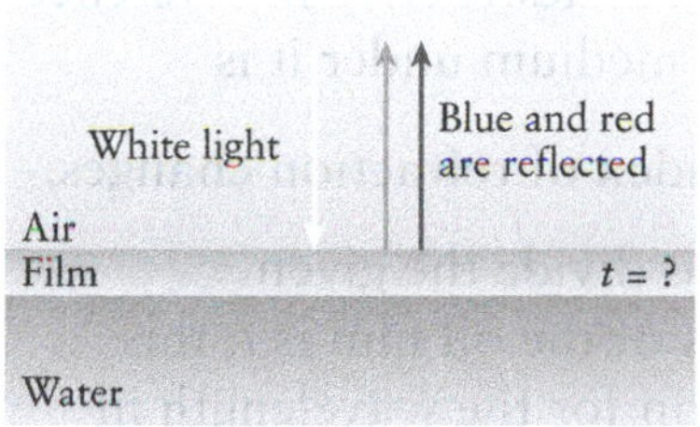

White light	Blue and red are reflected
Air	
Film	$t = ?$
Water	

Figure 23-7 Problem 63

SOLVE

Blue light:

$$2t = \left(m_b + \frac{1}{2} \right) \frac{\lambda_b}{n_{film}}$$

Red light:

$$2t = \left(m_r + \frac{1}{2} \right) \frac{\lambda_r}{n_{film}}$$

Solving for m's:

$$\left(m_b + \frac{1}{2} \right) \frac{\lambda_b}{n_{film}} = \left(m_r + \frac{1}{2} \right) \frac{\lambda_r}{n_{film}}$$

$$\left(m_b + \frac{1}{2} \right) (500 \text{ nm}) = \left(m_r + \frac{1}{2} \right) (700 \text{ nm})$$

$$5m_b + \frac{5}{2} = 7m_r + \frac{7}{2}$$

$$5m_b - 7m_r = 1$$

This is true if $m_b = 3$ and $m_r = 2$.

Thickness:

$$t = \left(m_\text{b} + \frac{1}{2}\right)\frac{\lambda_\text{b}}{2n_\text{film}} = \left(3 + \frac{1}{2}\right)\frac{500 \text{ nm}}{2(1.35)} = \boxed{648 \text{ nm}}$$

REFLECT

Since we were looking for the minimum thickness of the film, we chose the smallest values of m_b and m_r that satisfied the equation $5m_\text{b} - 7m_\text{r} = 1$. The next set of integers that satisfies that relationship is $m_\text{b} = 10$ and $m_\text{r} = 7$, which corresponds to $t = 1944$ nm.

Get Help: Interactive Example – Film on Water
P'Cast 23.8 – Reducing the Reflection

23.67

SET UP

A pool of water ($n_\text{water} = 1.33$) covered with a thin film of oil ($t = 450$ nm, $n_\text{oil} = 1.45$) is illuminated with white light and viewed from straight above. We want to know which visible wavelengths are *not* present in the reflected light, which means these wavelengths undergo total destructive interference. The condition on the path difference for destructive interference in the case where the film has a higher index of refraction than the medium under it is $(m + 1)\dfrac{\lambda}{n_\text{oil}}$. Note that the wavelength of light changes when the index of refraction changes. We are interested in the wavelength of light in the oil, so we need to divide the given wavelength by the index of refraction of the oil. Since the thickness of the oil film is t, this path difference should be equal to $2t$. We can solve for an expression for the wavelength in terms of the integer m. This will allow us to plug in consecutive values for m (starting from $m = 0$) and determine which wavelengths lie in the visible region (around 380 nm–750 nm)

SOLVE

$$2t = (m + 1)\frac{\lambda}{n_\text{oil}}$$

$$\lambda = \frac{2tn_\text{oil}}{m + 1} = \frac{2(450 \text{ nm})(1.45)}{m + 1} = \frac{1305 \text{ nm}}{m + 1}$$

For $m = 0$, $\lambda = 1305$ nm, which is in the infrared.

For $m = 1$, $\boxed{\lambda = 653 \text{ nm}}$, which is red/orange.

For $m = 2$, $\boxed{\lambda = 435 \text{ nm}}$, which is violet/indigo.

For $m = 3$, $\lambda = 326$ nm, which is ultraviolet.

REFLECT

The visible wavelength that is strongly reflected is 522 nm (green).

Get Help: Interactive Example – Film on Water
P'Cast 23.8 – Reducing the Reflection

23.73

SET UP

The beam from a He-Ne laser illuminates a single slit of width $w = 1850$ nm. In the resulting diffraction pattern, the first dark fringe appears at an angle of 20.0 degrees from the central maximum. We can use the expression for the angular position of the first dark fringe, $\sin(\theta) = \dfrac{\lambda}{w}$, to calculate the wavelength of the laser light.

SOLVE

$$\sin(\theta) = \frac{\lambda}{w}$$

$$\lambda = w\sin(\theta) = (1850 \text{ nm})\sin(20.0°) = \boxed{633 \text{ nm}}$$

REFLECT

A typical He-Ne laser appears red, and a wavelength of 633 nm is well within the range of red visible light.

23.79

SET UP

Light of wavelength λ is sent through a circular aperture of diameter D. The diffraction pattern is projected on a screen located a distance $L = 0.85$ m from the aperture. The first dark ring is 15,000λ from the center of the central maximum. The angular position of the first dark ring is given by $\sin(\theta) = 1.22\dfrac{\lambda}{D}$. Using the small angle approximation, we can relate this to $\tan(\theta)$ by geometry. Putting all of this together will allow us to calculate the diameter of the aperture D.

SOLVE

Geometry:

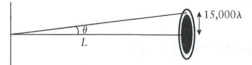

Figure 23-8 Problem 79

$$\tan(\theta) = \frac{15,000\lambda}{L}$$

Diameter of aperture:

$$\sin(\theta) = 1.22\frac{\lambda}{D}$$

Using the small angle approximation, $\sin(\theta) \approx \tan(\theta) = \dfrac{15,000\lambda}{L}$:

$$\frac{15{,}000\lambda}{L} = 1.22\frac{\lambda}{D}$$

$$D = \frac{1.22L}{15{,}000} = \frac{1.22(0.85 \text{ m})}{15{,}000} = \boxed{6.91 \times 10^{-5} \text{ m}}$$

REFLECT

The wavelength of visible light is on the order of hundreds of nanometers. If we assume 5×10^{-7} m as a typical value, the first dark ring is 7.5×10^{-3} m from the center of the central maximum. This is approximately 2 orders of magnitude smaller than the distance between the aperture and the screen, so our use of the small angle approximation is justified.

Get Help: P'Cast 23.9 – The Hubble Space Telescope

23.87

SET UP

A flat glass surface ($n_{\text{glass}} = 1.54$) has a uniform layer of water ($n_{\text{water}} = 1.33$) on top of the glass. We want to know the minimum angle of incidence that light coming from the glass must strike the glass-water interface such that the light is totally internally reflected by the water-air interface. The easiest way of tackling this problem is by working backwards. First, we can calculate the critical angle for the water-air interfaces from $\theta_c = \arcsin\left(\dfrac{n_{\text{air}}}{n_{\text{water}}}\right)$. This angle is also equal to the angle at which the light is refracted by the glass-water interface, which means we can apply Snell's law and calculate the incident angle of the light in the glass.

SOLVE

Critical angle for water-air interface:

$$\theta_c = \arcsin\left(\frac{n_{\text{air}}}{n_{\text{water}}}\right) = \arcsin\left(\frac{1.00}{1.33}\right) = 48.75°$$

Angle of incidence for the glass-water interface:

$$n_{\text{glass}} \sin(\theta_1) = n_{\text{water}} \sin(\theta_c)$$

$$\theta_1 = \arcsin\left(\frac{n_{\text{water}}}{n_{\text{glass}}} \sin(\theta_c)\right) = \arcsin\left(\frac{1.33}{1.54} \sin(48.75°)\right) = \boxed{40.5°}$$

REFLECT

We could have saved ourselves an intermediate step by realizing the water layer effectively does not enter into the calculation. Applying Snell's law at the first interface, we find $n_{\text{glass}}\sin(\theta_1) = n_{\text{water}}\sin(\theta_2)$. Applying it again at the second interface we get $n_{\text{water}}\sin(\theta_2) = n_{\text{air}}\sin(90°)$. Combining these two expressions, we can eliminate the term related to the water and see that $n_{\text{glass}}\sin(\theta_1) = n_{\text{air}}$, or

$$\theta_1 = \arcsin\left(\frac{n_{\text{air}}}{n_{\text{glass}}}\right) = \arcsin\left(\frac{1.00}{1.54}\right) = 40.5°.$$

23.93

SET UP

Unpolarized light with an intensity of $I_0 = 100\dfrac{\text{W}}{\text{m}^2}$ is incident on three polarizers. Two of them are placed with their transmission axes perpendicular to each other. A third polarizer is placed in between the two crossed polarizers such that the transmission axis of the second polarizer is oriented 30 degrees relative to that of the first. This means the transmission axis of the third polarizer is oriented 60 degrees relative to that of the second. In general, the intensity of unpolarized light drops by a factor of two when it passes through a linear polarizer. The intensity of the now linearly polarized light after it passes through a polarizer depends on the previous intensity and the square of the cosine of the angle between the polarization axis of the light and the transmission axis of the polarizer. The orientation of the middle polarizer that maximizes the transmitted intensity can be found by differentiating the general form for the transmitted intensity with respect to θ, setting it equal to zero, and solving for θ. Since the first and third polarizers are perpendicular to one another, we can call the angle between polarizers 1 and 2 θ and the angle between polarizers 2 and 3 $(90° - \theta)$.

SOLVE

Part a)

$$I_1 = \frac{I_0}{2}$$

$$I_2 = I_1 \cos^2(30°) = \left(\frac{I_0}{2}\right)\left(\frac{\sqrt{3}}{2}\right)^2 = \frac{3I_0}{8}$$

$$I_3 = I_2 \cos^2(60°) = \left(\frac{3I_0}{8}\right)\left(\frac{1}{2}\right)^2 = \frac{3I_0}{32} = \frac{3}{32}\left(100\frac{\text{W}}{\text{m}^2}\right) = \boxed{9.375\frac{\text{W}}{\text{m}^2}}$$

Part b)

Finding I_3 in terms of the general angle θ:

$$I_2 = I_1 \cos^2(\theta) = \frac{I_0}{2}\cos^2(\theta)$$

$$I_3 = I_2 \cos^2(90° - \theta) = \left(\frac{I_0}{2}\cos^2(\theta)\right)(\sin^2(\theta)) = \frac{I_0}{2}(\sin(\theta)\cos(\theta))^2$$

$$= \frac{I_0}{2}\left(\frac{\sin(2\theta)}{2}\right)^2 = \frac{I_0}{8}\sin^2(2\theta)$$

Finding the maximum:

$$\frac{dI_3}{d\theta} = 0$$

$$\frac{d}{d\theta}\left[\frac{I_0}{8}\sin^2(2\theta)\right] = \frac{I_0}{8}(4\sin(2\theta)\cos(2\theta)) = \frac{I_0}{4}(\sin(4\theta)) = 0$$

$$4\theta = \arcsin(0) = 0, \pi, 2\pi...$$

The solution $\theta = 0$ yields a minimum intensity, which means the maximum occurs at

$$4\theta = \pi$$

$$\theta = \frac{\pi}{4} = \boxed{45°}$$

REFLECT
After the fact, it makes sense that the middle sheet should be at an angle of 45 degrees relative to each polarizer due to symmetry in order to maximize the transmitted intensity.

Get Help: Interactive Example – Polarization I
Interactive Example – Polarization II

23.99

SET UP
A slip of paper of thickness T is placed between the edges of two thin plates of glass that have a length of $L = 0.125$ m. This causes the top plate to make an angle θ with respect to the bottom plate. When light ($\lambda = 600 \times 10^{-9}$ m) is shone normally on the glass plates, interference fringes due to destructive interference are observed. The spacing along the plate between neighboring fringes is $x = 0.200 \times 10^{-3}$ m. The thickness of the air film changes along the length of the plates. We can use the relationship for destructive interference in order to calculate the difference in the thickness of the film at successive fringes; this distance is related to x by $\sin(\theta)$. Since we expect the thickness of one sheet of paper to be very small, the angle between the glass plates will also be very small, which means we can invoke the small angle approximation: $\tan(\theta) \approx \sin(\theta) \approx \theta$. This angle is also related to the thickness of the paper and the length of the bottom plate by $\tan(\theta)$, which is approximately equal to θ. By setting the two expressions for θ equal to one another, we can solve for T.

SOLVE

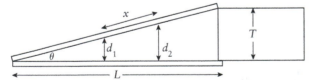

Figure 23-9 Problem 99

Destructive interference at successive fringes:

$$2d_1 = \left(m_1 + \frac{1}{2} \right)\lambda$$

$$2d_2 = \left(m_2 + \frac{1}{2} \right)\lambda = \left((m_1 + 1) + \frac{1}{2} \right)\lambda = \left(m_1 + \frac{3}{2} \right)\lambda$$

Horizontal distance between successive fringes:

$$2d_2 - 2d_1 = \left(m_1 + \frac{3}{2} \right)\lambda - \left(m_1 + \frac{1}{2} \right)\lambda = \lambda$$

$$d_2 - d_1 = \frac{\lambda}{2}$$

Angle in terms of the distance along the plate between successive fringes:

$$\sin(\theta) = \frac{d_2 - d_1}{x} = \frac{\left(\frac{\lambda}{2}\right)}{x} = \frac{\lambda}{2x}$$

$$\sin(\theta) \approx \theta \approx \frac{\lambda}{2x}$$

Angle in terms of the thickness of the paper:

$$\tan(\theta) = \frac{T}{L}$$

$$\tan(\theta) \approx \theta \approx \frac{T}{L}$$

Thickness of the paper:

$$\frac{T}{L} = \frac{\lambda}{2x}$$

$$T = \frac{\lambda L}{2x} = \frac{(600 \times 10^{-9} \text{ m})(0.125 \text{ m})}{2(0.200 \times 10^{-3} \text{ m})} = \boxed{1.88 \times 10^{-4} \text{ m} = 188 \ \mu m}$$

REFLECT

A thickness of about 0.2 mm seems reasonable for a single sheet of paper.

23.103

SET UP

The distance between the crest and adjacent trough of a tsunami wave is 250 mi, which means the wavelength will be twice this distance. The period of the tsunami wave is $T = 1$ hr. The speed of the wave is equal to the wavelength divided by the period. The time it takes the tsunami to travel a distance of $\Delta x = 600$ mi is equal to that distance divided by the wave speed. These waves travel between a 100-mile-wide opening. To determine whether we can use the diffraction relationship $w\sin(\theta) = m\lambda$, we need to compare the slit width w to the wavelength λ. The slit needs to be "narrow," which means w must be approximately the same size or smaller than the wavelength.

SOLVE

Part a)

$$v = \lambda f = \frac{\lambda}{T} = \frac{2(250 \text{ mi})}{1 \text{ hr}} = \boxed{500 \text{ mph}}$$

Part b)

$$v = \frac{\Delta x}{\Delta t}$$

$$\Delta t = \frac{\Delta x}{v} = \frac{600 \text{ mi}}{500 \text{ mph}} = \boxed{1.2 \text{ hr}}$$

Part c) The formula $w\sin(\theta) = m\lambda$ applies only if the wavelength is smaller than the slit width w. In this case $w = 100$ mi and $\lambda = 500$ mi, so the formula $\boxed{\text{would } \textit{not} \text{ apply}}$.

REFLECT
If we were to apply $w\sin(\theta) = m\lambda$ to find the first diffraction minimum, we would find that $\sin(\theta) = \dfrac{\lambda}{w} = \dfrac{500 \text{ mi}}{100 \text{ mi}} = 5$. The sine of an angle can only yield a result between -1 and 1, so there is no solution to this equation. Physically, this means the first diffraction minimum does not exist, and, thus, diffraction does not occur.

23.109

SET UP
An optical telescope can achieve an angular resolution of 0.25 arcsec $\left(1 \text{ arcsec} = \left(\dfrac{1}{3600}\right)^{\circ} = 4.85 \times 10^{-6} \text{ rad}\right)$. Rayleigh's criterion for the angular resolution through a circular aperture is $\sin(\theta_R) = 1.22\dfrac{\lambda}{D}$, where we're interested in the wavelengths of visible light. We will use a wavelength of 550×10^{-9} m to represent visible light. We can use this, along with the Rayleigh criterion and the desired resolution of 0.25 arcsec, to calculate the minimum value of D.

SOLVE

Part a)

$$\sin(\theta_R) = 1.22\frac{\lambda}{D}$$

$$D = 1.22\frac{\lambda}{\sin(\theta_R)} = 1.22\frac{\lambda}{\sin\left(\dfrac{1}{4}\text{arcsec} \times \dfrac{4.85 \times 10^{-6} \text{ rad}}{1 \text{ arcsec}}\right)} = (1.01 \times 10^6)\lambda$$

$$= (1.01 \times 10^6)(550 \times 10^{-9} \text{ m}) = 0.55 \text{ m} = 55 \text{ cm}$$

The minimum diameter aperture should be around $\boxed{55 \text{ cm}}$ to achieve a resolution of 0.25 arcsec.

Part b) Building a telescope with a diameter much larger than this won't improve the resolution significantly as long as you have to look through Earth's atmosphere. Telescopes are built larger than the diffraction-limited diameter for greater light collecting power, which allows you to see dimmer objects.

REFLECT
The Rayleigh criterion is just one rule of thumb used to determine whether or not two objects can be resolved through a circular aperture.

Get Help: P'Cast 23.9 – The Hubble Space Telescope

Chapter 24
Geometrical Optics

Conceptual Questions

24.1 Actually, a plane mirror does neither, but instead inverts objects back to front. If the mirror inverted right and left, then the object's right hand that points east would appear on the images as a right hand pointing toward the west. Because the image is inverted back to front, the object facing north is transformed into an image that faces south. Also, the object's right hand is transformed into a left hand in the image.

24.7 Additional information is needed. In accordance with the lensmaker's equation, if the radius of curvature of the front surface is larger, it is a diverging lens; if the radius of curvature of the front surface is smaller, it is a converging lens.

Get Help: Picture It: Converging Lens

Multiple-Choice Questions

24.15 B (The height of the image stays the same and the image distance increases). For a plane mirror, the image distance equals the object distance, so the image distance will increase as the object distance increases. Also, the image height is equal to the object height; since the height of the object doesn't change, neither will the height of the image.

24.19 B (real and inverted).

$$\frac{1}{d_O} + \frac{1}{d_I} = \frac{1}{f}$$

$$\frac{1}{d_I} = \frac{1}{f} - \frac{1}{d_O} = \frac{1}{\left(\frac{r}{2}\right)} - \frac{1}{r} = \frac{2}{r} - \frac{1}{r} = \frac{1}{r}$$

$$d_I = r$$

$$m = -\frac{d_I}{d_O} = -\frac{r}{r} = -1$$

A positive image distance means the image is real. A negative magnification indicates the image is inverted.

24.23 B (The objective lens is a short focal length, convex lens and the eyepiece functions as a simple magnifier). The real image from the objective lens of a compound microscope converges just inside the focal point of the eyepiece, which means the angular size of the overall image will be much larger than the angular size of the object.

Get Help: Interactive Example – Two Lens System
P'Cast 24.6 – Magnification

Estimation/Numerical Questions

24.27 A shiny spoon (f about 5 cm), a shiny salad bowl (f about 15 cm), and a reflector in a flashlight (f about 2 cm).

Get Help: Interactive Example – Concave Mirror

Problems

24.31

SET UP

Two flat mirrors are perpendicular to each other. An incoming beam of light makes an angle of 30 degrees with respect to the first mirror. The beam will reflect off the mirror at an angle of 30 degrees according to the law of reflection. The beam then makes an angle of 60 degrees with respect to the second mirror, or 30 degrees with respect to the normal of the second mirror. Again, applying the law of reflection, we see that the outgoing beam will make an angle of 30 degrees with respect to the normal of the second mirror.

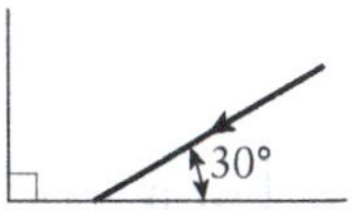

Figure 24-1 Problem 31

SOLVE

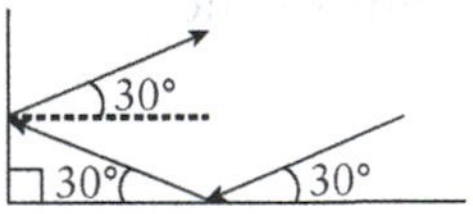

Figure 24-2 Problem 31

The outgoing beam will make an angle of $\boxed{30 \text{ degrees}}$ with respect to the normal of the second mirror.

REFLECT

Recognizing common geometric patterns will go a long way in solving optics problems. For example, the path of the ray leaving the first mirror and hitting the second mirror makes a 30-60-90 triangle with the two mirrors.

24.35

SET UP

A plane mirror is 10 m away from and parallel to a second plane mirror. An object is placed 3 m to the right of the left mirror, which means it is 7 m to the left of the right mirror. The original object creates an image in both the left and the right mirror. These images then act as objects for the other mirror and are thus reflected again. This process repeats and there will be an infinite number of images. In every case, the image distance is equal to the object distance for the mirror.

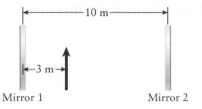

Figure 24-3 Problem 35

SOLVE

Illustration of the resulting images:

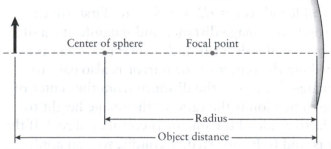

Figure 24-4 Problem 35

The first images formed in the left mirror are 3 m, 17 m, 23 m, 37 m, and 43 m to the left of the left mirror .

The first images formed in the right mirror are 7 m, 13 m, 27 m, 47 m, and 53 m to the left of the left mirror .

REFLECT

You can see this effect for yourself at the hair salon or in a mirror maze or even if you have two mirrors in your bathroom.

24.41

SET UP

We can use ray tracing in order to determine if there are any situations where a real image is formed by a spherical, concave mirror.

SOLVE

An arrow is placed in front of a spherical, concave mirror. We will trace two light rays from the tip of the arrow to the mirror and as they reflect from it.

Figure 24-5 Problem 41

The image of the point of the arrow forms where the two rays cross. The image is inverted relative to the object.

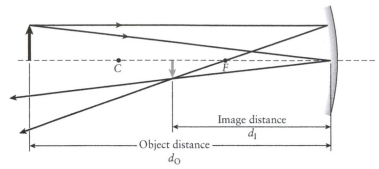

Figure 24-6 Problem 41

Two light rays are traced from the tip of the arrow in order to find the location of the image.

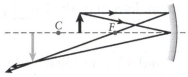

Figure 24-7 Problem 41

As long as $d_O > f$, the image will be real in a concave, spherical mirror. In this case, the focal point is located between the mirror and the object.

REFLECT
An object placed outside the center of curvature of a concave, spherical mirror will result in an image smaller than the object; an object placed in between the center of curvature and the focal point of a concave, spherical mirror will result in an image larger than the object.

Get Help: Picture It – Spherical Mirror

24.43

SET UP
An object is placed in front of a concave mirror that has a radius of curvature of $r = 10$ cm. The object distance is $d_O = 8.0$ cm, and the focal length is $f = r/2 = +5.0$ cm. First, we can draw a ray trace diagram of the setup to determine the image distance and magnification of the image. A parallel ray from the tip of the object will reflect off of the mirror through the focal point. A ray from the tip of the object striking the center of the mirror is also easy to reflect by applying the law of reflection. The image distance is the distance from the center of the mirror to the position of the image; the magnification is the ratio of the image height to the object height. An image located on the reflective side of the mirror is considered real. If the image is upside down relative to the object, it is said to be inverted. Secondly, we can apply the mirror equation, $\dfrac{1}{d_O} + \dfrac{1}{d_I} = \dfrac{1}{f}$, and its sign conventions to confirm the results from the ray trace diagram. A positive image distance is considered real, and a negative magnification corresponds to an inverted image.

SOLVE

Part a)

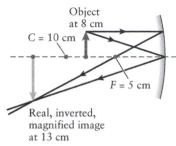

Figure 24-8 Problem 43

Part b)

Image distance:

$$\frac{1}{d_O} + \frac{1}{d_I} = \frac{1}{f}$$

$$\frac{1}{d_I} = \frac{1}{f} - \frac{1}{d_O} = \frac{1}{5.0 \text{ cm}} - \frac{1}{8.0 \text{ cm}} = \frac{3.0}{40 \text{ cm}}$$

$$d_I = \frac{40 \text{ cm}}{3.0} = \boxed{13 \text{ cm}}$$

Magnification:

$$m = -\frac{d_I}{d_O} = -\frac{13 \text{ cm}}{8.0 \text{ cm}} = \boxed{-1.7}$$

The image is $\boxed{\text{real}}$ because the image distance is positive. The image is $\boxed{\text{inverted}}$ because $m < 0$.

REFLECT

A ray trace diagram must be drawn to scale if you want to quantitatively measure distances and heights.

Get Help: Interactive Example – Concave Mirror
P'Cast 24.4 – An Object Far from a Concave Mirror
P'Cast 24.5 – An Object Close to a Concave Mirror

24.47

SET UP

The radius of curvature of a spherical concave mirror is 15 cm, which means its focal length is $f = +7.5$ cm. An object ($h_O = 20$ cm) is positioned at three different object distances: $d_I = +10$ cm, $d_I = +20$ cm, and $d_I = +100$ cm. We can use the mirror equation, $\frac{1}{d_O} + \frac{1}{d_I} = \frac{1}{f}$, and its sign conventions to calculate the image distance and the image height, as well as determine whether the image is real or virtual, upright or inverted. As a reminder, a positive image distance is considered real, and a negative image height corresponds to an inverted image.

SOLVE

Image distance in general:

$$\frac{1}{d_O} + \frac{1}{d_I} = \frac{1}{f}$$

$$\frac{1}{d_I} = \frac{1}{f} - \frac{1}{d_O} = \frac{d_O - f}{f d_O}$$

$$d_I = \frac{f d_O}{d_O - f}$$

Image height in general:

$$m = \frac{h_I}{h_O} = -\frac{d_I}{d_O}$$

$$h_I = -\frac{d_I}{d_O} h_O$$

Part a)

Image distance:

$$d_I = \frac{f d_O}{d_O - f} = \frac{(7.5 \text{ cm})(10 \text{ cm})}{(10 \text{ cm}) - (7.5 \text{ cm})} = \boxed{30 \text{ cm}}$$

The image is $\boxed{\text{real}}$ because $d_I > 0$.

Image height:

$$h_I = -\frac{d_I}{d_O} h_O = -\left(\frac{30 \text{ cm}}{10 \text{ cm}}\right)(20 \text{ cm}) = -60 \text{ cm}$$

The image is $\boxed{\text{60 cm tall and inverted}}$ because $h_I < 0$.

Part b)

Image distance:

$$d_I = \frac{f d_O}{d_O - f} = \frac{(7.5 \text{ cm})(20 \text{ cm})}{(20 \text{ cm}) - (7.5 \text{ cm})} = \boxed{12 \text{ cm}}$$

The image is $\boxed{\text{real}}$ because $d_I > 0$.

Image height:

$$h_I = -\frac{d_I}{d_O} h_O = -\left(\frac{12 \text{ cm}}{20 \text{ cm}}\right)(20 \text{ cm}) = -12 \text{ cm}$$

The image is $\boxed{\text{12 cm tall and inverted}}$ because $h_I < 0$.

Part c)

Image distance:

$$d_I = \frac{f d_O}{d_O - f} = \frac{(7.5 \text{ cm})(100 \text{ cm})}{(100 \text{ cm}) - (7.5 \text{ cm})} = \boxed{8.11 \text{ cm}}$$

The image is $\boxed{\text{real}}$ because $d_I > 0$.

Image height:

$$h_I = -\frac{d_I}{d_O}h_O = -\left(\frac{8.11 \text{ cm}}{100 \text{ cm}}\right)(20 \text{ cm}) = -1.62 \text{ cm}$$

The image is $\boxed{1.62 \text{ cm tall and inverted}}$ because $h_I < 0$.

REFLECT

A real object placed at a position outside of the focal point of a spherical concave mirror will always give a real image:

$$d_I = \frac{fd_O}{d_O - f} > 0, \text{ if } (d_O - f) > 0$$

Get Help: Interactive Example – Concave Mirror
P'Cast 24.4 – An Object Far from a Concave Mirror
P'Cast 24.5 – An Object Close to a Concave Mirror

24.55

SET UP

The radius of curvature of a spherical convex mirror is 20 cm, which means its focal length is $f = -10$ cm. An object ($h_O = 10$ cm) is positioned at three different object distances: $d_I = +20$ cm, $d_I = +50$ cm, and $d_I = +100$ cm. We can use the mirror equation, $\frac{1}{d_O} + \frac{1}{d_I} = \frac{1}{f}$, and its sign conventions to calculate the image distance and the image height, as well as determine whether the image is real or virtual, upright or inverted. As a reminder, a positive image distance is considered real, and a negative image height corresponds to an inverted image.

SOLVE

Image distance in general:

$$\frac{1}{d_O} + \frac{1}{d_I} = \frac{1}{f}$$

$$\frac{1}{d_I} = \frac{1}{f} - \frac{1}{d_O} = \frac{d_O - f}{fd_O}$$

$$d_I = \frac{fd_O}{d_O - f}$$

Image height in general:

$$m = \frac{h_I}{h_O} = -\frac{d_I}{d_O}$$

$$h_I = -\frac{d_I}{d_O}h_O$$

Part a)

Image distance:

$$d_I = \frac{fd_O}{d_O - f} = \frac{(-10 \text{ cm})(20 \text{ cm})}{(20 \text{ cm}) - (-10 \text{ cm})} = \boxed{-6.67 \text{ cm}}$$

The image is $\boxed{\text{virtual}}$ because $d_I < 0$.

Image height:

$$h_I = -\frac{d_I}{d_O}h_O = -\left(\frac{-6.67 \text{ cm}}{20 \text{ cm}}\right)(10 \text{ cm}) = 3.33 \text{ cm}$$

The image is $\boxed{3.33 \text{ cm tall and upright}}$ because $h_I > 0$.

Part b)

Image distance:

$$d_I = \frac{fd_O}{d_O - f} = \frac{(-10 \text{ cm})(50 \text{ cm})}{(50 \text{ cm}) - (-10 \text{ cm})} = \boxed{-8.33 \text{ cm}}$$

The image is $\boxed{\text{virtual}}$ because $d_I < 0$.

Image height:

$$h_I = -\frac{d_I}{d_O}h_O = -\left(\frac{-8.33 \text{ cm}}{50 \text{ cm}}\right)(10 \text{ cm}) = 1.67 \text{ cm}$$

The image is $\boxed{1.67 \text{ cm tall and upright}}$ because $h_I > 0$.

Part c)

Image distance:

$$d_I = \frac{fd_O}{d_O - f} = \frac{(-10 \text{ cm})(100 \text{ cm})}{(100 \text{ cm}) - (-10 \text{ cm})} = \boxed{-9.09 \text{ cm}}$$

The image is $\boxed{\text{virtual}}$ because $d_I < 0$.

Image height:

$$h_I = -\frac{d_I}{d_O}h_O = -\left(\frac{-9.09 \text{ cm}}{100 \text{ cm}}\right)(10 \text{ cm}) = 0.909 \text{ cm}$$

The image is $\boxed{0.909 \text{ cm tall and upright}}$ because $h_I > 0$.

REFLECT

A real object placed *anywhere* in front of a spherical convex mirror will always give a virtual image, *i.e.*, $d_I = \frac{fd_O}{d_O - f} < 0$. The term $(d_O - f)$ will always be positive for a real object and a convex mirror.

Get Help: P'Cast 24.4 – An Object Far from a Convex Mirror
P'Cast 24.5 – An Object Close to a Convex Mirror

24.61

SET UP

In order to prove that all images produced by a spherical convex mirror are virtual, we need to show mathematically that the image distance is negative regardless of our choice of d_O. We can rearrange the mirror equation, solve for the image distance, and apply the necessary sign conventions, mainly that a spherical convex mirror has a negative focal length.

SOLVE

$$\frac{1}{d_O} + \frac{1}{d_I} = \frac{1}{f}$$

$$\frac{1}{d_I} = \frac{1}{f} - \frac{1}{d_O} = \frac{d_O - f}{f d_O}$$

$$d_I = \frac{f d_O}{d_O - f}$$

The focal length of a spherical convex mirror is always negative, which means the term $(d_O - f)$ will always be positive for a real object. Therefore,

$$d_I = \frac{f d_O}{d_O - f} < 0$$

REFLECT

A negative image distance means the image is virtual.

Get Help: P'Cast 24.4 – An Object Far from a Convex Mirror

P'Cast 24.5 – An Object Close to a Convex Mirror

24.65

SET UP

A real image created by reflection in a spherical mirror appears in front of the mirrored surface. A real image created by refraction through a lens appears behind the lens.

SOLVE

$\boxed{\text{No}}$, a real image in a converging lens occurs on the opposite side of the lens from the object.

REFLECT

We expect light to be reflected by a mirror and transmitted by a lens.

24.69

SET UP

The power of a converging lens is 5 diopters, which means its focal length is $f = \dfrac{1}{P} = \dfrac{1}{5 \text{ m}^{-1}} = 0.20$ m $= +20$ cm. An object ($h_O = 10$ cm) is positioned at three different object distances: $d_I = +5$ cm, $d_I = +10$ cm, and $d_I = +50$ cm. We can use the lens equation, $\dfrac{1}{d_O} + \dfrac{1}{d_I} = \dfrac{1}{f}$, and its sign conventions to calculate the image distance and the image height, as well as determine whether the image is real or virtual, upright or inverted. As a reminder, a positive image distance is considered real, and a negative image height corresponds to an inverted image. We can also draw a ray trace diagram of each situation to confirm our numerical results.

SOLVE

Image distance in general:

$$\frac{1}{d_O} + \frac{1}{d_I} = \frac{1}{f}$$

$$\frac{1}{d_I} = \frac{1}{f} - \frac{1}{d_O} = \frac{d_O - f}{f d_O}$$

$$d_I = \frac{f d_O}{d_O - f}$$

Image height in general:

$$m = \frac{h_I}{h_O} = -\frac{d_I}{d_O}$$

$$h_I = -\frac{d_I}{d_O} h_O$$

Part a)

Image distance:

$$d_I = \frac{f d_O}{d_O - f} = \frac{(20 \text{ cm})(5 \text{ cm})}{(5 \text{ cm}) - (20 \text{ cm})} = \boxed{-6.7 \text{ cm}}$$

The image is $\boxed{\text{virtual}}$ because $d_I < 0$.

Image height:

$$h_I = -\frac{d_I}{d_O} h_O = -\left(\frac{-6.7 \text{ cm}}{5 \text{ cm}}\right)(10 \text{ cm}) = 13 \text{ cm}$$

The image is $\boxed{\text{13 cm tall and upright}}$ because $h_I > 0$.

Ray trace diagram:

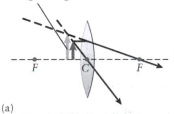

(a)

Figure 24-9 Problem 69

Part b)

Image distance:

$$d_I = \frac{f d_O}{d_O - f} = \frac{(20 \text{ cm})(10 \text{ cm})}{(10 \text{ cm}) - (20 \text{ cm})} = \boxed{-20 \text{ cm}}$$

The image is $\boxed{\text{virtual}}$ because $d_I < 0$.

Image height:

$$h_I = -\frac{d_I}{d_O} h_O = -\left(\frac{-20 \text{ cm}}{10 \text{ cm}}\right)(10 \text{ cm}) = 20 \text{ cm}$$

The image is $\boxed{\text{20 cm tall and upright}}$ because $h_I > 0$.

Ray trace diagram:

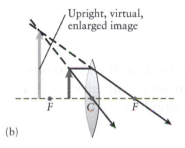

(b)

Figure 24-10 Problem 69

Part c)

Image distance:

$$d_I = \frac{f d_O}{d_O - f} = \frac{(20 \text{ cm})(20 \text{ cm})}{(20 \text{ cm}) - (20 \text{ cm})} = \infty$$

$\boxed{\text{No image is formed}}$.

Ray trace diagram:

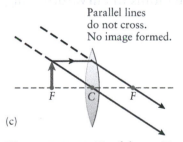

(c)

Figure 24-11 Problem 69

Part d)

Image distance:

$$d_I = \frac{f d_O}{d_O - f} = \frac{(20 \text{ cm})(50 \text{ cm})}{(50 \text{ cm}) - (20 \text{ cm})} = \boxed{33 \text{ cm}}$$

The image is $\boxed{\text{real}}$ because $d_I > 0$.

Image height:

$$h_I = -\frac{d_I}{d_O}h_O = -\left(\frac{33 \text{ cm}}{50 \text{ cm}}\right)(10 \text{ cm}) = -6.7 \text{ cm}$$

The image is $\boxed{6.7 \text{ cm tall and inverted}}$ because $h_I < 0$.

Ray trace diagram:

Inverted, real, reduced image

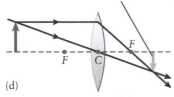

(d)

Figure 24-12 Problem 69

REFLECT

An object placed at the focal point of a lens will not create an image because the rays come out parallel and do not converge. This is the reverse situation of an object at infinity emitting parallel rays, which are the converged at the focal point by a converging lens.

Get Help: Picture It – Converging Lens

24.73

SET UP

A plano-concave, glass lens ($n = 1.60$) has a focal length of $f = -31.8$ cm. We can calculate the radius of curvature of the concave surface from the lensmaker's equation, $\frac{1}{f} = (n-1)\left(\frac{1}{R_1} - \frac{1}{R_2}\right)$. We'll assume that the front surface of the lens is planar, which means $R_1 = \infty$. For the lensmaker's equation, R_2 is positive for a concave surface and negative for a convex surface. Once we know the R_2 for a concave surface, we can invert its sign in order to calculate the focal length of a plano-convex, glass lens with the same radius of curvature using the lensmaker's equation.

SOLVE

Part a)

$$\frac{1}{f} = (n-1)\left(\frac{1}{R_1} - \frac{1}{R_2}\right)$$

$$\frac{1}{f(n-1)} = \frac{1}{R_1} - \frac{1}{R_2} = \frac{1}{\infty} - \frac{1}{R_2} = -\frac{1}{R_2}$$

$$R_2 = -f(n-1) = -(-31.8 \text{ cm})((1.60) - 1) = \boxed{19.1 \text{ cm} = 0.191 \text{ m}}$$

Part b)

$$\frac{1}{f} = (n-1)\left(\frac{1}{R_1} - \frac{1}{R_2}\right)$$

$$f = \frac{1}{(n-1)\left(\dfrac{1}{R_1} - \dfrac{1}{R_2}\right)} = \frac{1}{(n-1)\left(\dfrac{1}{\infty} - \dfrac{1}{R_2}\right)} = \frac{1}{(n-1)\left(-\dfrac{1}{R_2}\right)}$$

$$= \frac{-R_2}{n-1} = \frac{-(-19.1 \text{ cm})}{(1.60) - 1} = \boxed{+31.8 \text{ cm}}$$

REFLECT
It makes sense that changing the back surface of the lens from concave to convex should convert the lens from a diverging one to a converging one. Also, since we're not changing the magnitude of the radius of curvature of that surface, the magnitude of the focal length should not change.

24.81

SET UP
A microscope that consists of an objective lens and an eyepiece separated by $L = 16$ cm has a total magnification of $m_{total} = 400$. The focal length of the objective is $f_o = 0.60$ cm. The total magnification of a system of lenses is equal to the product of the magnifications due to each individual lens; in this case, $m_{total} = m_o m_e$. A microscope is constructed in such a way that a specimen located at the focal plane of the objective lens creates an image near the focal plane of the eyepiece. Since the distance between the lenses is much larger than the focal length of the lenses, the magnification of the objective lens is equal to the distance separating the two lenses divided by the focal length of the objective lens. The eyepiece takes the image from the objective as its object and creates an image at the near point of your eye ($N = 25$ cm), which means the magnification due to the eyepiece is equal to N divided by the focal length of the eyepiece. By rearranging the expression for the total magnification, we can solve for the focal length of the eyepiece.

SOLVE

$$m_{total} = m_o m_e = \left(\frac{L}{f_o}\right)\left(\frac{N}{f_e}\right)$$

$$f_e = \frac{LN}{f_o m_{total}} = \frac{(16 \text{ cm})(25 \text{ cm})}{(0.60 \text{ cm})(400)} = \boxed{1.7 \text{ cm}}$$

REFLECT
In a compound microscope consisting of two lenses, the focal length of the objective lens is always smaller than the focal length of the eyepiece.

Get Help: Interactive Example – Two Lens System
P'Cast 24.6 – Magnification

24.85

SET UP

A very small, thin plano-concave lens ($f_A = -45.0$ cm) has the same principal axis as a concave mirror with a radius of curvature of $+60.0$ cm. The focal length of a spherical mirror is equal to the radius of curvature divided in half, so $f_B = 30.0$ cm. An object is placed $d_{O,A} = 15.0$ cm in front of the lens. We can use the thin lens equation to determine the location of the image produced by the lens. This image now acts as the object for the concave mirror, so we can use the mirror equation to calculate the location of the image after it was reflected by the mirror. This will be the final image because the lens is so much smaller than the mirror; essentially none of the light rays reflected by the mirror will pass through the lens. If the image distance from the mirror calculation is positive, then the image is real; if it's negative, the image is virtual. Whether or not the image is upright or inverted can be found by calculating the total magnification due to both the lens and the mirror. A convex mirror of the same magnitude radius of curvature will have a focal length of $f_B = -30.0$ cm. We can follow the same process as the concave mirror to determine where the image is produced by this convex mirror.

SOLVE

Part a)

Image formed by the lens:

$$\frac{1}{d_{O,A}} + \frac{1}{d_{I,A}} = \frac{1}{f_A}$$

$$\frac{1}{d_{I,A}} = \frac{1}{f_A} - \frac{1}{d_{O,A}} = \frac{d_{O,A} - f_A}{f_A d_{O,A}}$$

$$d_{I,A} = \frac{f_A d_{O,A}}{d_{O,A} - f_A} = \frac{(-45.0 \text{ cm})(15.0 \text{ cm})}{(15.0 \text{ cm}) - (-45.0 \text{ cm})} = -11.25 \text{ cm}$$

Image formed by the mirror:

The image formed by the lens is located 11.25 cm in front of the lens, which means $d_{O,B} = (11.25 \text{ cm}) + (20.0 \text{ cm}) = 31.25$ cm.

$$\frac{1}{d_{O,B}} + \frac{1}{d_{I,B}} = \frac{1}{f_B}$$

$$\frac{1}{d_{I,B}} = \frac{1}{f_B} - \frac{1}{d_{O,B}} = \frac{d_{O,B} - f_B}{f_B d_{O,B}}$$

$$d_{I,B} = \frac{f_B d_{O,B}}{d_{O,B} - f_B} = \frac{(30.0 \text{ cm})(31.25 \text{ cm})}{(31.25 \text{ cm}) - (30.0 \text{ cm})} = 750 \text{ cm}$$

The image formed by the mirror is located $\boxed{750 \text{ cm in front of the mirror}}$ (or 730 cm in front of the lens).

Part b)

The final image is $\boxed{\text{real because } d_{I,B} > 0}$.

Magnification:

$$m_{\text{total}} = m_A m_B = \left(-\frac{d_{I,A}}{d_{O,A}}\right)\left(-\frac{d_{I,B}}{d_{O,B}}\right) = \left(\frac{-11.25 \text{ cm}}{15.0 \text{ cm}}\right)\left(\frac{750 \text{ cm}}{31.25 \text{ cm}}\right) = -18$$

The image is $\boxed{\text{inverted because } m_{\text{total}} < 0}$.

Part c)

The only thing that changes is that $f_B = -30.0$ cm.

Therefore,

$$d_{I,B} = \frac{f_B d_{O,B}}{d_{O,B} - f_B} = \frac{(-30.0 \text{ cm})(31.25 \text{ cm})}{(31.25 \text{ cm}) - (-30.0 \text{ cm})} = -15.3 \text{ cm}$$

The image created by the mirror is $\boxed{\text{virtual because } d_{I,B} < 0}$.

Magnification:

$$m_{\text{total}} = m_A m_B = \left(-\frac{d_{I,A}}{d_{O,A}}\right)\left(-\frac{d_{I,B}}{d_{O,B}}\right) = \left(\frac{-11.25 \text{ cm}}{15.0 \text{ cm}}\right)\left(\frac{-15.3 \text{ cm}}{31.25 \text{ cm}}\right) = 0.367$$

The image is $\boxed{\text{upright because } m_{\text{total}} > 0}$.

REFLECT

If the lens were the same size as the mirror, the light reflected by the mirror will pass through the lens again, in the opposite direction as part a. The real image created by the concave mirror will act as a virtual object for the lens in this case.

Get Help: P'Cast 24.6 – Magnification

24.91

SET UP

The *tapetum lucidum* is a highly reflective membrane just behind the retina of the eyes of cats, which have a typical diameter of $d = 1.25$ cm. We can model this membrane as a concave spherical mirror with a radius of curvature of $r = d/2 = +0.625$ cm. We will assume that the light entering the cat's eye is traveling parallel to the principal axis of the lens. Parallel incoming light rays will be focuses at the focal point; the focal length f for a concave spherical mirror is equal to $r/2$. The eye is filled with fluid with a refractive index of $n = 1.4$. Although this will affect the refraction of the light rays, it has no effect on reflection.

SOLVE

Part a)

$$f = \frac{r}{2} = \frac{0.625 \text{ cm}}{2} = 0.31 \text{ cm} = 3.1 \text{ mm}$$

The light will be focused $\boxed{3.1 \text{ mm in front of the retina}}$.

Part b) Reflection is not affected by the index of refraction, so the answer would be the same as in part (a).

REFLECT

We could have also calculated the image distance from the mirror equation:

$$\frac{1}{d_O} + \frac{1}{d_I} = \frac{1}{\infty} + \frac{1}{d_I} = \frac{1}{f}$$

$$d_I = f$$

24.95

SET UP

A specific person's near point was measured to be 15.0 cm and his far point is 2.75 m. These points for normal vision are 25.0 cm and infinity, respectively. It's not considered an issue if a person's near point is "too close"; his unaided eye can see things closer than an average person can, which is fine. Therefore, he wouldn't need to correct this issue and would only require lenses with a single focal length. The corrective contact lenses would need to take an object located at infinity and produce an image at the person's uncorrected far point. Contact lenses are located on a person's eye, so the image distance would be equal to the far point distance. Plugging all of this information with the correct sign conventions into the thin-lens equation will allow us to calculate the required focal length of the contact lenses. The power of the contact lenses is equal to the reciprocal of the focal length.

SOLVE

Part a) The person's far point is too close, so he needs a single focal length lens to correct this issue.

Part b)

$$\frac{1}{d_O} + \frac{1}{d_I} = \frac{1}{f}$$

$$\frac{1}{\infty} + \frac{1}{d_I} = \frac{1}{f}$$

$$f = d_I = \boxed{-2.75 \text{ m}}$$

Part c)

$$P = \frac{1}{f} = \frac{1}{-2.75 \text{ m}} = \boxed{-0.364 \text{ diopters}}$$

REFLECT

People who have trouble seeing things off in the distance are nearsighted because they can only see things near to them.

24.97

SET UP

A zoom lens for a digital camera is first zoomed such that the focal length is $f = 200$ mm $= 0.200$ m. The camera is focused on an object that is a distance $d_O = 15.0$ m from the lens. We can use the lens equation to calculate the distance between the lens and the photosensor array inside the camera, *i.e.*, the image distance d_I. The width of the image is essentially the same as the image height, h_I, which we can find from the definition of the magnification. The focal length of the zoom lens is then changed to $f = 18$ mm; since we already know what the image distance must be, we can calculate the closest an object can be for this case.

SOLVE

Part a)

$$\frac{1}{d_O} + \frac{1}{d_I} = \frac{1}{f}$$

$$\frac{1}{d_I} = \frac{1}{f} - \frac{1}{d_O} = \frac{d_O - f}{fd_O}$$

$$d_I = \frac{fd_O}{d_O - f} = \frac{(0.200 \text{ m})(15.0 \text{ m})}{(15.0 \text{ m}) - (0.200 \text{ m})} = \boxed{0.203 \text{ m} = 203 \text{ mm}}$$

Part b)

$$m = \frac{h_I}{h_O} = -\frac{d_I}{d_O}$$

$$h_I = -\frac{d_I}{d_O}h_O = -\left(\frac{0.203 \text{ m}}{15.0 \text{ m}}\right)(38 \text{ cm}) = \boxed{0.51 \text{ cm}}$$

Part c)

$$\frac{1}{d_O} + \frac{1}{d_I} = \frac{1}{f}$$

$$\frac{1}{d_O} = \frac{1}{f} - \frac{1}{d_I} = \frac{d_I - f}{fd_I}$$

$$d_O = \frac{fd_I}{d_I - f} = \frac{(18 \text{ mm})(52 \text{ mm})}{(52 \text{ mm}) - (18 \text{ mm})} = \boxed{28 \text{ mm}}$$

REFLECT
The image distance must remain constant since it is dictated by the dimensions and construction of the camera.

Chapter 25
Relativity

Conceptual Questions

25.3 A frame of reference or reference frame is a coordinate system with respect to which we will make observations or measurements. An inertial frame is one that moves at constant speed relative to another, that is, we refer to a frame of reference attached to a nonaccelerating object as an inertial frame.

25.9 If the object is moving relative to you, you measure its length by finding the difference between the coordinates of its end points at the same time.

Multiple-Choice Questions

25.21 D ($1.00c$). The speed of light is constant in all reference frames.

25.25 A (the clock in Colorado runs slow). The effect of the Earth's gravitational field decreases in magnitude as you move away from the Earth.

Estimation/Numerical Questions

25.31 About 5% of the world population.

Problems

25.35

SET UP

A frame of reference, S, is fixed on the surface of the Earth with the x-axis pointing toward the east, y-axis pointing toward the north, and the z-axis pointing up. A second frame of reference, S', is moving at a constant speed of $V = 4$ m/s to the east. We can use the Galilean transformation to mathematically describe the relationships between x and x', y and y', and z and z'. Once we have the expressions for x', y', and z', we can calculate their values given $t = 4$ s, $x = 2$ m, $y = 1$ m, and $z = 0$ m.

SOLVE

Part a)

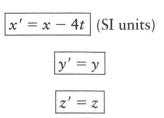

$$\boxed{x' = x - 4t}\;\text{(SI units)}$$

$$\boxed{y' = y}$$

$$\boxed{z' = z}$$

Part b)

$$x' = x - 4t = (2) - 4(4) = \boxed{-14 \text{ m}}$$

$$y' = y = \boxed{1 \text{ m}}$$

$$z' = z = \boxed{0 \text{ m}}$$

REFLECT

The second reference frame was moving toward the east $(+x)$, so the sign of V should be positive.

25.41

SET UP

An airplane lands in a river during a hurricane. We will use a coordinate system where positive x points east and positive y points north. The velocity of the airplane relative to the wind is $\vec{v}_{aw} = \left(30\dfrac{\text{m}}{\text{s}}\right)\hat{x}$; the velocity of the wind relative to the ground is $\vec{v}_{wg} = \left(20\dfrac{\text{m}}{\text{s}}\right)\hat{y}$; and the velocity of the river relative to the ground is $\vec{v}_{rg} = -\left(5\dfrac{\text{m}}{\text{s}}\right)\hat{y}$. The velocity of the airplane relative to the river, $\vec{v}_{ar}$, is equal to the vector sum $\vec{v}_{ar} = \vec{v}_{aw} + \vec{v}_{wg} + \vec{v}_{gr}$. Recall that the sign of the velocity changes when the subscripts are reversed, for example, $\vec{v}_{gr} = -\vec{v}_{rg}$.

SOLVE

$$\vec{v}_{aw} = \left(30\frac{\text{m}}{\text{s}}\right)\hat{x}$$

$$\vec{v}_{wg} = \left(20\frac{\text{m}}{\text{s}}\right)\hat{y}$$

$$\vec{v}_{rg} = -\left(5\frac{\text{m}}{\text{s}}\right)\hat{y}$$

$$\vec{v}_{ar} = \vec{v}_{aw} + \vec{v}_{wg} + \vec{v}_{gr} = \vec{v}_{aw} + \vec{v}_{wg} - \vec{v}_{rg} = \left(30\frac{\text{m}}{\text{s}}\right)\hat{x} + \left(20\frac{\text{m}}{\text{s}}\right)\hat{y} - \left(-5\frac{\text{m}}{\text{s}}\right)\hat{y}$$

$$= \left(30\frac{\text{m}}{\text{s}}\right)\hat{x} + \left(25\frac{\text{m}}{\text{s}}\right)\hat{y}$$

Magnitude:

$$v_{ar} = \sqrt{v_{ar,\,x}^2 + v_{ar,\,y}^2} = \sqrt{\left(30\frac{\text{m}}{\text{s}}\right)^2 + \left(25\frac{\text{m}}{\text{s}}\right)^2} = \boxed{39.1\frac{\text{m}}{\text{s}}}$$

Direction:

$$\theta = \arctan\left(\frac{v_{ar,\,y}}{v_{ar,\,x}}\right) = \arctan\left(\frac{\left(25\frac{\text{m}}{\text{s}}\right)}{\left(30\frac{\text{m}}{\text{s}}\right)}\right) = \boxed{39.8° \text{ north of east}}$$

REFLECT

We could have just as easily reported our answer in vector form, $\vec{v}_{ar} = \left(30\frac{m}{s}\right)\hat{x} + \left(25\frac{m}{s}\right)\hat{y}$.

A good mnemonic for determining the correct order when calculating a relative velocity is that the leftmost and rightmost subscripts should be the ones found in your answer and that all neighboring, intermediate subscripts should be the same. For example, in this problem we want our answer to have the subscripts $\vec{v}_{ar}$, so the leftmost subscript should be "a" and the rightmost should be "r." Then, we can fill in the rest of the sum by matching the subscripts: $\vec{v}_{ar} = \vec{v}_{aw} + \vec{v}_{wg} + \vec{v}_{gr}$.

25.49

SET UP

The average lifetime of muons traveling at a speed of $V = 0.98c$ was measured to be $\Delta t = 11\ \mu s$. If the muons were at rest, then their average lifetime would be equal to the proper time $\Delta t_{proper} = \Delta t\sqrt{1 - \dfrac{V^2}{c^2}}$.

SOLVE

$$\Delta t_{proper} = \Delta t\sqrt{1 - \frac{V^2}{c^2}} = (11\ \mu s)\sqrt{1 - \frac{(0.98c)^2}{c^2}} = (11\ \mu s)\sqrt{1 - (0.98)^2} = \boxed{2.19\ \mu s}$$

REFLECT

The proper time is a time measurement made at rest with respect to the reference frame. We expect the lifetime measured for the moving muons to be longer than the ones at rest due to time dilation.

Get Help: P'Cast 25.3 – Moving Clock

25.55

SET UP

A car, with a proper length of $L_{proper} = 3.20$ m, is moving in the x direction in the reference frame S. The reference frame S' moves at a speed of $V = 0.80c$ toward positive x. The length of the car according to observers in S' is equal to $L = \dfrac{1}{\gamma}L_{proper}$, where $\gamma = \dfrac{1}{\sqrt{1 - \dfrac{V^2}{c^2}}}$.

SOLVE

$$L = \frac{1}{\gamma}L_{proper} = \left(\sqrt{1 - \frac{V^2}{c^2}}\right)L_{proper} = \left(\sqrt{1 - \frac{(0.80c)^2}{c^2}}\right)(3.20\text{ m})$$

$$= (\sqrt{1 - (0.80)^2})(3.20\text{ m}) = \boxed{1.92\text{ m}}$$

REFLECT

The observed length of the car in the moving reference frame should be smaller than the proper length of the car due to length contraction.

Get Help: P'Cast 25.5 – Meter Stick
P'Cast 25.8 – A Galactic Competition
P'Cast 25.9 – A Galactic Competition, Reconsidered

25.59

SET UP

We need to determine the speed of a pion that travels $D_{proper} = 100$ m before it decays. The average lifetime, at rest, of a pion is $t_{proper} = 2.60 \times 10^{-8}$ s. The speed we're interested in is equal to the distance in the frame of the pion D divided by the lifetime of the pion t_{proper}. Since the pions are moving, the distance they travel will be contracted by a factor of $\gamma = \dfrac{1}{\sqrt{1 - \dfrac{V^2}{c^2}}}$. We can then solve for V to calculate the speed of the pions.

SOLVE

$$V = \frac{D}{t_{proper}} = \frac{\left(\frac{1}{\gamma}D_{proper}\right)}{t_{proper}} = \frac{D_{proper}}{\gamma t_{proper}}$$

$$\gamma V = \frac{D_{proper}}{t_{proper}}$$

$$\frac{\gamma V}{c} = \frac{D_{proper}}{t_{proper}c}$$

$$\left(\frac{\gamma V}{c}\right)^2 = \left(\frac{D_{proper}}{t_{proper}c}\right)^2$$

$$\frac{\left(\frac{V}{c}\right)^2}{\left(1 - \left(\frac{V}{c}\right)^2\right)} = \left(\frac{D_{proper}}{t_{proper}c}\right)^2$$

$$\left(\frac{V}{c}\right)^2 = \frac{\left(\frac{D_{proper}}{t_{proper}c}\right)^2}{1 + \left(\frac{D_{proper}}{t_{proper}c}\right)^2}$$

$$\frac{V}{c} = \sqrt{\frac{\left(\frac{D_{proper}}{t_{proper}c}\right)^2}{1 + \left(\frac{D_{proper}}{t_{proper}c}\right)^2}} = \sqrt{\frac{\left(\frac{100 \text{ m}}{(2.60 \times 10^{-8} \text{ s})\left(3.00 \times 10^8 \frac{\text{m}}{\text{s}}\right)}\right)^2}{1 + \left(\frac{100 \text{ m}}{(2.60 \times 10^{-8} \text{ s})\left(3.00 \times 10^8 \frac{\text{m}}{\text{s}}\right)}\right)^2}} = \boxed{0.997}$$

$$V = 0.997c = 0.997\left(3.00 \times 10^8 \frac{\text{m}}{\text{s}}\right) = \boxed{2.99 \times 10^8 \frac{\text{m}}{\text{s}}}$$

REFLECT

This solution is equivalent to transforming the measurement of the average lifetime of the pions in the moving frame, $V = \dfrac{D_{proper}}{t} = \dfrac{D_{proper}}{\gamma t_{proper}}$.

Get Help: P'Cast 25.5 – Meter Stick

P'Cast 25.8 – A Galactic Competition

P'Cast 25.9 – A Galactic Competition, Reconsidered

25.63

SET UP

A spaceship is traveling past Earth when it fires a rocket in the backward direction relative to the spaceship, and we want to know the velocity of the rocket relative to Earth. We will attach a stationary reference frame S to the Earth and reference frame S' to the spaceship. The spaceship is traveling past Earth at a speed of $V = 0.92c$. The speed of the rocket relative to the spaceship is $v'_x = -0.75c$; the negative sign means the rocket is shot backward relative to the spaceship. We can rearrange the expression for the Lorentz transformation (Equation 25-21) in order to solve for the velocity of the rocket relative to the Earth v_x.

SOLVE

$$v'_x = \frac{v_x - V}{1 - \left(\dfrac{V}{c^2}v_x\right)}$$

$$v'_x\left(1 - \left(\frac{V}{c^2}v_x\right)\right) = v_x - V$$

$$v'_x + V = v_x + \left(\frac{V}{c^2}v_x\right)v'_x$$

$$v_x = \frac{v'_x + V}{1 + \left(\dfrac{V}{c^2}v_x\right)} = \frac{(-0.75c) + (0.92c)}{1 + \left(\dfrac{(0.92c)}{c^2}(-0.75c)\right)} = \frac{0.17c}{1 - 0.69} = \boxed{0.548c}$$

REFLECT

It makes sense that a rocket shot backward from a spaceship should be moving at a slower speed relative to the Earth than the speed of the rocket relative to the Earth.

25.67

SET UP

An electron (rest mass $m_e = 9.11 \times 10^{-31}$ kg) travels at $v = 0.444c$. The magnitude of the relativistic momentum is $p = \gamma m_0 v$, where $\gamma = \dfrac{1}{\sqrt{1 - \left(\dfrac{v}{c}\right)^2}}$. The total energy E of the electron is the sum of its relativistic kinetic energy K and the energy not associated with its motion, *i.e.*, its rest mass energy, E_0. The rest energy is equal to $E_0 = m_0c^2$, and its total energy is equal to $E = \gamma m_0 c^2$.

SOLVE

Relativistic gamma:

$$\gamma = \frac{1}{\sqrt{1 - \left(\dfrac{v}{c}\right)^2}} = \frac{1}{\sqrt{1 - \left(\dfrac{0.444c}{c}\right)^2}} = \frac{1}{\sqrt{1 - 0.197}} = \frac{1}{\sqrt{0.803}} = 1.116$$

Part a)

$$p = \gamma m_0 v = \gamma m_e (0.444c)$$

$$= (1.116)(9.11 \times 10^{-31} \text{ kg})(0.444)\left(3.00 \times 10^8 \frac{\text{m}}{\text{s}}\right) = \boxed{1.35 \times 10^{-22} \frac{\text{kg} \cdot \text{m}}{\text{s}}}$$

Part b)

$$K + E_0 = E$$

$$K = E - E_0 = \gamma m_0 c^2 - m_0 c^2 = (\gamma - 1)m_0 c^2 = ((1.116) - 1)m_e c^2$$

$$= (0.116)(9.11 \times 10^{-31} \text{ kg})\left(3.00 \times 10^8 \frac{\text{m}}{\text{s}}\right)^2 = \boxed{9.51 \times 10^{-15} \text{ J}}$$

Part c)

$$E_0 = m_0 c^2 = m_e c^2 = (9.11 \times 10^{-31} \text{ kg})\left(3.00 \times 10^8 \frac{\text{m}}{\text{s}^2}\right)^2 = \boxed{8.20 \times 10^{-14} \text{ J}}$$

Part d)

$$E = \gamma m_0 c^2 = (1.116)m_e c^2 = (1.116)(9.11 \times 10^{-31} \text{ kg})\left(3.00 \times 10^8 \frac{\text{m}}{\text{s}^2}\right)^2 = \boxed{9.15 \times 10^{-14} \text{ J}}$$

REFLECT

The work done on an object required to increase the speed approaches infinity since relativistic gamma approaches infinity as the object's speed approaches c.

Get Help: P'Cast 25.11 – Rest Energy

25.73

SET UP

A proton has a rest energy of $E_0 = m_0 c^2 = 1.50 \times 10^{-10}$ J and a momentum of $p = \gamma m_0 v = 1.067 \times 10^{-19}\frac{\text{kg} \cdot \text{m}}{\text{s}}$. In order to find the speed of the proton, we first need to derive an expression for the relativistic energy of the proton in terms of its momentum. This will allow us to easily solve for γ (actually, γ^2) in terms of our known quantities, the rest energy and the momentum. The expressions for the rest energy, momentum, and relativistic energy are $E_0 = m_0 c^2$, $p = \gamma m_0 v$, and $E = mc^2 = \gamma E_0$, respectively. Finally, using the definition of γ we can solve for the speed of the proton.

SOLVE

Total energy in terms of momentum:

$$E = mc^2 = \gamma m_0 c^2 = \frac{m_0 c^2}{\sqrt{1 - \left(\frac{v}{c}\right)^2}}$$

$$m^2 c^4 = \frac{m_0^2 c^4}{1 - \left(\dfrac{v}{c}\right)^2}$$

$$m^2 c^4 - m^2 v^2 c^2 = m_0^2 c^4$$

$$m^2 c^4 = E^2 = m^2 v^2 c^2 + m_0^2 c^4 = (\gamma m_0)^2 v^2 c^2 + m_0^2 c^4 = p^2 c^2 + m_0^2 c^4$$

$$E^2 = p^2 c^2 + m_0^2 c^4$$

Solving for gamma:

$$E^2 = (\gamma m_0 c^2)^2 = p^2 c^2 + m_0^2 c^4$$

$$\gamma^2 = \frac{p^2 c^2 + (m_0 c^2)^2}{(m_0 c^2)^2}$$

$$= \frac{\left(1.067 \times 10^{-19} \dfrac{\text{kg} \cdot \text{m}}{\text{s}}\right)^2 \left(3.00 \times 10^8 \dfrac{\text{m}}{\text{s}}\right)^2 + (1.50 \times 10^{-10} \text{ J})^2}{(1.50 \times 10^{-10} \text{ J})^2} = 1.0455$$

Solving for the speed:

$$\gamma^2 = \frac{1}{1 - \left(\dfrac{v}{c}\right)^2}$$

$$\left(\frac{v}{c}\right)^2 = \frac{\gamma^2 - 1}{\gamma^2}$$

$$v = c\sqrt{\frac{\gamma^2 - 1}{\gamma^2}} = c\sqrt{\frac{1.0455 - 1}{1.0455}} = \boxed{0.209c}$$

REFLECT

We couldn't have simply divided the momentum by γm_0 to solve for v because γ is also a function of v.

Get Help: P'Cast 25.11 – Rest Energy

25.75

SET UP

An elevator near the Earth's surface is accelerating downward at 18 m/s². We can use the general theory of relativity to determine the free-fall acceleration an observer inside the elevator would experience.

SOLVE

$$a = \left(-9.8 \frac{\text{m}}{\text{s}^2}\right) - \left(-18 \frac{\text{m}}{\text{s}^2}\right) = 8.2 \frac{\text{m}}{\text{s}^2}$$

The observer in the elevator would measure the free-fall acceleration as $\boxed{8.2 \text{ m/s}^2 \text{ in the upward direction}}$.

REFLECT

We can do our own thought experiment to confirm our answer. The observer inside the elevator throws a ball horizontally. If the elevator car were stationary in free space, the ball would go in a straight line until it hit the wall. If the elevator car were accelerating downward, the ball would hit the wall at a location *higher* than the previous case because the entire elevator would have shifted downward due to its acceleration.

25.77

SET UP

Observers in a reference frame S see one explosion at $x_1 = 580$ m and then a second explosion $\Delta t = 4.5$ μs later at $x_2 = 1500$ m. Reference frame S' is moving along the positive x-axis at a speed v. We can calculate v from the data measured by those in S. Observers in S' see the explosions occur at the same point in space. Since the observers in S' see the explosions occur at the same point in space, we know that the proper time is the time separation measured in S'. The time between explosions as measured in S' is related to the time interval measured in S through relativistic gamma, $\Delta t = \gamma \Delta t_{proper}$.

SOLVE

Speed

$$v = \frac{x_2 - x_1}{\Delta t} = \frac{(1500 \text{ m}) - (580 \text{ m})}{4.5 \times 10^{-6} \text{ s}} = 2.044 \times 10^8 \frac{\text{m}}{\text{s}}$$

Time separation

$$\Delta t = \gamma \Delta t_{proper}$$

$$\Delta t_{proper} = \frac{\Delta t}{\gamma} = \Delta t \sqrt{1 - \left(\frac{v}{c}\right)^2} = (4.5 \text{ }\mu\text{s}) \sqrt{1 - \left(\frac{\left(2.044 \times 10^8 \frac{\text{m}}{\text{s}}\right)}{\left(3.00 \times 10^8 \frac{\text{m}}{\text{s}}\right)}\right)^2} = \boxed{3.29 \text{ }\mu\text{s}}$$

REFLECT

The proper time interval should be smaller than the time interval observed by those in S due to the effects of time dilation.

25.83

SET UP

A jet plane flies at $v = 300$ m/s relative to an observer on the ground. There is a clock aboard the plane, as well as one on the ground; the two were synchronized at the start. We want to know the distance (as measured by the observer on the ground) the plane must fly such that the clock on the plane is 10 s behind the clock on the ground. This time difference is the difference between the dilated time experienced on the plane and the proper time experienced on the ground. From this, we can calculate the proper time; the distance the plane must fly is equal to the plane's speed relative to the ground multiplied by the proper time. When calculating relativistic gamma, it will be helpful to use the approximation for $(1 - x)^{-\frac{1}{2}}$ when x is much smaller than 1: $(1 - x)^{-\frac{1}{2}} \approx 1 + \frac{1}{2}x$.

SOLVE

Relativistic gamma:

$$\gamma = \frac{1}{\sqrt{1 - \left(\frac{v}{c}\right)^2}} = \frac{1}{\sqrt{1 - \left(\frac{\left(300\frac{m}{s}\right)}{\left(3.00 \times 10^8 \frac{m}{s}\right)}\right)^2}} = \frac{1}{\sqrt{1 - (1 \times 10^{-12})}} = [1 - (1 \times 10^{-12})]^{-\frac{1}{2}}$$

$$\approx 1 + \frac{1}{2}(1 \times 10^{-12})$$

Time observed on the ground:

$$t - t_{\text{proper}} = \gamma t_{\text{proper}} - t_{\text{proper}} = (\gamma - 1)t_{\text{proper}} = 10 \text{ s}$$

$$t_{\text{proper}} = \frac{10 \text{ s}}{(\gamma - 1)}$$

Distance of the flight as measured by an observer on the ground:

$$d = v t_{\text{proper}} = v\left(\frac{10 \text{ s}}{\gamma - 1}\right) \approx \left(300\frac{m}{s}\right)\frac{(10 \text{ s})}{\left(1 + \frac{1}{2}(1 \times 10^{-12})\right) - 1} = \left(300\frac{m}{s}\right)\frac{2(10 \text{ s})}{1 \times 10^{-12}}$$

$$= \boxed{6 \times 10^{15} \text{ m}}$$

REFLECT

This is approximately 0.6 light years. For comparison, the distance between the Earth and the Sun is about 1.5×10^{11} m; the distance the jet has to travel is 40,000 times larger than that.

25.87

SET UP

In one month, a household uses 411 kWh of electrical energy and 201 therms of gas heating (1.0 therm = 29.3 kWh). After converting these values into joules, we can use $E = mc^2$ to calculate the mass necessary to meet the monthly energy needs for this house.

SOLVE

Conversions:

$$411 \text{ kWh} \times \frac{3600 \text{ s}}{1 \text{ hr}} \times \frac{1000 \text{ W}}{1 \text{ kW}} = 1.480 \times 10^9 \text{ J}$$

$$201 \text{ therms} \times \frac{29.3 \text{ kWh}}{1.0 \text{ therm}} \times \frac{3600 \text{ s}}{1 \text{ hr}} \times \frac{1000 \text{ W}}{1 \text{ kW}} = 2.120 \times 10^{10} \text{ J}$$

Total energy:

$$E = (2.120 \times 10^{10} \text{ J}) + (1.480 \times 10^9 \text{ J}) = 2.268 \times 10^{10} \text{ J}$$

Mass:

$$E = mc^2$$

$$m = \frac{E}{c^2} = \frac{2.268 \times 10^{10} \text{ J}}{\left(3.00 \times 10^8 \frac{\text{m}}{\text{s}}\right)^2} = 2.52 \times 10^{-7} \text{ kg} \times \frac{10^6 \text{ mg}}{1 \text{ kg}} = \boxed{0.252 \text{ mg}}$$

REFLECT

This would be the mass necessary if we were harvesting the energy directly from the atoms that make up the molecules.

Get Help: P'Cast 25.11 – Rest Energy
P'Cast 25.12 – The Sun

25.95

SET UP

A 25-year-old captain pilots a spaceship to a planet that is $\Delta x = 42$ ly from Earth. The captain needs to be no more than 60 years old when she arrives at the planet, which means the proper time interval, as observed by the spaceship, is $\Delta t_{proper} = 35$ y. The speed of the spaceship relative to the Earth v is equal to Δx divided by the time interval as observed by the Earth Δt, where $\Delta t = \gamma \Delta t_{proper}$. Using the definition of relativistic gamma, $\gamma = \dfrac{1}{\sqrt{1 - \left(\dfrac{v}{c}\right)^2}}$, we can solve

for v. In order to make the math easier, we'll use the fact that 1 ly = (1 y)c. The captain sends a radio signal back to Earth as soon as it reaches the planet. The total time after the launch when the signal arrives is equal to the time it takes the ship to reach the planet plus the time it takes the signal to travel back to Earth. We can relate these times to the distance between Earth and the planet as well as the speeds of the spaceship and the speed of light.

SOLVE

Part a)

$$\Delta t = \gamma \Delta t_{proper}$$

$$v = \frac{\Delta x}{\Delta t} = \frac{\Delta x}{\gamma \Delta t_{proper}} = \frac{\Delta x}{\Delta t_{proper}} \sqrt{1 - \left(\frac{v}{c}\right)^2}$$

$$v^2 = \left(\frac{\Delta x}{\Delta t_{proper}}\right)^2 - \left(\frac{\Delta x}{\Delta t_{proper}}\right)^2 \left(\frac{v}{c}\right)^2$$

$$v = \frac{\Delta x}{\Delta t_{proper}\sqrt{1 + \left(\dfrac{\Delta x}{c\Delta t_{proper}}\right)^2}} = \frac{c\Delta x}{\sqrt{(c\Delta t_{proper})^2 + (\Delta x)^2}} = \frac{c(42 \text{ ly})}{\sqrt{(c(35 \text{ y}))^2 + (42 \text{ ly})^2}}$$

$$= \frac{c((42 \text{ y})c)}{\sqrt{(c(35 \text{ y}))^2 + ((42 \text{ y})c)^2}} = \frac{c(42 \text{ y})}{\sqrt{(35 \text{ y})^2 + (42 \text{ y})^2}} = \boxed{0.77c}$$

Part b)

$$\Delta t_{total} = \Delta t_{spaceship} + \Delta t_{signal} = \frac{\Delta x}{v} + \frac{\Delta x}{c} = \Delta x\left(\frac{1}{v} + \frac{1}{c}\right) = \Delta x\left(\frac{1}{0.77c} + \frac{1}{c}\right) = \frac{\Delta x}{c}\left(\frac{1}{0.77} + 1\right)$$

$$= \frac{(42\ ly)}{c}(2.30) = \frac{(42\ y)c}{c}(2.30) = \boxed{96.7\ y}$$

REFLECT

A distance of 42 ly is around 4×10^{17} m!

Chapter 26
Modern and Atomic Physics

Conceptual Questions

26.5 The shortest wavelength is the wavelength that is emitted when a free electron is captured into the lowest energy state, which is 91.2 nm.

26.13 The electron has the longer wavelength.

26.17 Because the rest mass of a macroscopic object is so large, the de Broglie wavelength is too small to observe. The wavelength of everyday objects is many orders of magnitude less than the radius of an atom. This is far too small for us to observe.

Multiple-Choice Questions

26.21 A (a photon of ultraviolet radiation). Out of all of the choices, ultraviolet light has the highest frequency and, thus, the highest energy.

26.25 C (decreases). The increase in wavelength in the Compton effect as a function of wavelength is $\Delta\lambda \propto (1 - \cos(\theta))$. The energy is inversely proportional to the wavelength.

Get Help: Picture It – Compton Scattering

Estimation/Numerical Questions

26.29 Your body radiates about 500–1000 W

26.35 An electron must travel at 7.3×10^6 m/s for its de Broglie wavelength to be about the size of an atom (about 10^{-10} m).

Get Help: P'Cast 26.6 – A Slow Hummingbird

Problems

26.41

SET UP
A blackbody at $T = 300$ K radiates heat into its immediate surroundings. Wien's displacement law, $\lambda_{max} = \dfrac{0.290 \text{ K} \cdot \text{cm}}{T}$, tells us the wavelength at which the maximum intensity is emitted.

SOLVE

$$\lambda_{max} = \frac{0.290 \text{ K} \cdot \text{cm}}{300 \text{ K}} = \boxed{9.67 \times 10^{-4} \text{ cm} = 9670 \text{ nm}}$$

REFLECT

This corresponds to infrared light, which makes sense for a blackbody at a relatively low temperature.

26.45

SET UP

We are asked to calculate the range of frequencies and energies in the visible spectrum of light, which is approximately from $\lambda_{violet} = 380 \times 10^{-9}$ m to $\lambda_{red} = 750 \times 10^{-9}$ m. The frequency of a photon is inversely related to its wavelength, $f = \dfrac{c}{\lambda}$; the energy is directly proportional to the frequency, $E = hf$.

SOLVE

Frequencies:

$$f_{violet} = \frac{c}{\lambda_{violet}} = \frac{\left(3.00 \times 10^8 \dfrac{\text{m}}{\text{s}}\right)}{380 \times 10^{-9} \text{ m}} = \boxed{7.89 \times 10^{14} \text{ Hz}}$$

$$f_{red} = \frac{c}{\lambda_{red}} = \frac{\left(3.00 \times 10^8 \dfrac{\text{m}}{\text{s}}\right)}{750 \times 10^{-9} \text{ m}} = \boxed{4.00 \times 10^{14} \text{ Hz}}$$

Energies:

$$E_{violet} = hf_{violet} = (6.63 \times 10^{-34} \text{ J} \cdot \text{s})(7.89 \times 10^{14} \text{ Hz}) = 5.23 \times 10^{-19} \text{ J}$$

$$5.23 \times 10^{-19} \text{ J} \times \frac{1 \text{ eV}}{1.6 \times 10^{-19} \text{ J}} = \boxed{3.27 \text{ eV}}$$

$$E_{red} = hf_{red} = (6.63 \times 10^{-34} \text{ J} \cdot \text{s})(4.00 \times 10^{14} \text{ Hz}) = 2.65 \times 10^{-19} \text{ J}$$

$$2.65 \times 10^{-19} \text{ J} \times \frac{1 \text{ eV}}{1.6 \times 10^{-19} \text{ J}} = \boxed{1.66 \text{ eV}}$$

REFLECT

Keep in mind that the wavelength is inversely proportional to the energy; red light has a longer wavelength than violet light, but is less energetic. We could have also solved this in eV directly using $h = 4.1357 \times 10^{-15}$ eV $\cdot$ s.

26.51

SET UP

Light ($\lambda = 195 \times 10^{-9}$ m) strikes a metal surface and photoelectrons are produced moving with a maximum speed of $v_{max} = 0.004c$. The expression relating the maximum kinetic energy of the photoelectrons to the work function is $K_{max} = hf - \Phi_0$. The threshold wavelength for the metal corresponds to the wavelength absorbed when the kinetic energy of the photoelectrons is equal to zero. Once we know the numerical value of the work function, we can search the Internet for a list of work functions to determine which metal our sample may be.

SOLVE

Part a)

$$K_{max} = hf - \Phi_0$$

$$\Phi_0 = hf - K_{max} = hf - \frac{1}{2}m_e v_{max}^2 = \frac{hc}{\lambda} - \frac{1}{2}m_e(0.004c)^2$$

$$= \frac{(6.63 \times 10^{-34} \text{ J} \cdot \text{s})\left(3.00 \times 10^8 \frac{\text{m}}{\text{s}}\right)}{195 \times 10^{-9} \text{ m}} - \frac{1}{2}(9.11 \times 10^{-31} \text{ kg})(0.004)^2\left(3.00 \times 10^8 \frac{\text{m}}{\text{s}}\right)^2$$

$$= 3.63 \times 10^{-19} \text{ J} \times \frac{1 \text{ eV}}{1.6 \times 10^{-19} \text{ J}} = \boxed{2.27 \text{ eV}}$$

Part b)

$$K_{max} = 0 = hf_{min} - \Phi_0 = \frac{hc}{\lambda_{max}} - \Phi_0$$

$$\lambda_{max} = \frac{hc}{\Phi_0} = \frac{(6.63 \times 10^{-34} \text{ J} \cdot \text{s})\left(3.00 \times 10^8 \frac{\text{m}}{\text{s}}\right)}{3.63 \times 10^{-19} \text{ J}} = \boxed{5.46 \times 10^{-7} \text{ m} = 546 \text{ nm}}$$

Part c) The metal could be either sodium ($\Phi_0 = 2.28$ eV) or potassium ($\Phi_0 = 2.3$ eV).

REFLECT

We can't know the exact type of metal without more information and a more precise measurement.

Get Help: P'Cast 26.2 – Photoelectric Effect with Cesium

26.55

SET UP

The momentum of a photon is equal to the energy of the photon divided by the speed of light. By using the expression for the energy in terms of the wavelength of the light, we can easily calculate the momenta of photons with wavelengths of 550×10^{-9} m and 0.0711×10^{-9} m.

SOLVE

Relating momentum to wavelength:

$$p = \frac{E}{c} = \left(\frac{hc}{\lambda}\right)\left(\frac{1}{c}\right) = \frac{h}{\lambda}$$

Part a)

$$p = \frac{h}{\lambda} = \frac{6.63 \times 10^{-34} \text{ J} \cdot \text{s}}{550 \times 10^{-9} \text{ m}} = \boxed{1.21 \times 10^{-27} \frac{\text{kg} \cdot \text{m}}{\text{s}}}$$

Part b)

$$p = \frac{h}{\lambda} = \frac{6.63 \times 10^{-34} \text{ J} \cdot \text{s}}{0.0711 \times 10^{-9} \text{ m}} = \boxed{9.32 \times 10^{-24} \frac{\text{kg} \cdot \text{m}}{\text{s}}}$$

REFLECT

We can also use the fact that $hc = 1240$ eV $\cdot$ nm to represent the momentum in units of eV/c. Our answers in these units would be 2.25 eV/c and 17,400 eV/c, respectively.

Get Help: Picture It – Compton Scattering

26.61

SET UP

Photons with a wavelength of $\lambda_i = 0.1400$ nm are scattered at varying angles—0 degrees, 30 degrees, 45 degrees, 60 degrees, 90 degrees, and 180 degrees—off of carbon atoms. The shift in wavelength of the Compton scattered photons is given by $\Delta\lambda = \lambda_C(1 - \cos(\theta))$, where $\lambda_C = 0.00243$ nm; the wavelength of the scattered photons can be calculated from this relationship. The kinetic energy of the scattered electrons is equal to the difference between the energy of the initial photons and the energy of the scattered photons.

SOLVE

Wavelength of scattered photon:

$$\Delta\lambda = \lambda_f - \lambda_i = \lambda_C(1 - \cos(\theta))$$

$$\lambda_f = \lambda_i + \lambda_C(1 - \cos(\theta))$$

Kinetic energy of scattered electrons:

$$K = \frac{hc}{\lambda_i} - \frac{hc}{\lambda_f} = hc\left(\frac{1}{\lambda_i} - \frac{1}{\lambda_f}\right)$$

Part a)

$$\lambda_f = \lambda_i + \lambda_C(1 - \cos(\theta)) = (0.1400 \text{ nm}) + (0.00243 \text{ nm})(1 - \cos(0°)) = \boxed{0.1400 \text{ nm}}$$

$$K = hc\left(\frac{1}{\lambda_i} - \frac{1}{\lambda_f}\right) = (1240 \text{ eV} \cdot \text{nm})\left(\frac{1}{0.1400 \text{ nm}} - \frac{1}{0.1400 \text{ nm}}\right) = \boxed{0}$$

Part b)

$$\lambda_f = \lambda_i + \lambda_C(1 - \cos(\theta)) = (0.1400 \text{ nm}) + (0.00243 \text{ nm})(1 - \cos(30°)) = \boxed{0.14033 \text{ nm}}$$

$$K = hc\left(\frac{1}{\lambda_i} - \frac{1}{\lambda_f}\right) = (1240 \text{ eV} \cdot \text{nm})\left(\frac{1}{0.1400 \text{ nm}} - \frac{1}{0.14033 \text{ nm}}\right) = \boxed{20.8 \text{ eV}}$$

Part c)

$$\lambda_f = \lambda_i + \lambda_C(1 - \cos(\theta)) = (0.1400 \text{ nm}) + (0.00243 \text{ nm})(1 - \cos(45°)) = \boxed{0.14071 \text{ nm}}$$

$$K = hc\left(\frac{1}{\lambda_i} - \frac{1}{\lambda_f}\right) = (1240 \text{ eV} \cdot \text{nm})\left(\frac{1}{0.1400 \text{ nm}} - \frac{1}{0.14071 \text{ nm}}\right) = \boxed{44.7 \text{ eV}}$$

Part d)

$$\lambda_f = \lambda_i + \lambda_C(1 - \cos(\theta)) = (0.1400 \text{ nm}) + (0.00243 \text{ nm})(1 - \cos(60°)) = \boxed{0.14122 \text{ nm}}$$

$$K = hc\left(\frac{1}{\lambda_i} - \frac{1}{\lambda_f}\right) = (1240 \text{ eV} \cdot \text{nm})\left(\frac{1}{0.1400 \text{ nm}} - \frac{1}{0.14122 \text{ nm}}\right) = \boxed{75.3 \text{ eV}}$$

Part e)

$$\lambda_f = \lambda_i + \lambda_C(1 - \cos(\theta)) = (0.1400 \text{ nm}) + (0.00243 \text{ nm})(1 - \cos(90°)) = \boxed{0.14243 \text{ nm}}$$

$$K = hc\left(\frac{1}{\lambda_i} - \frac{1}{\lambda_f}\right) = (1240 \text{ eV} \cdot \text{nm})\left(\frac{1}{0.1400 \text{ nm}} - \frac{1}{0.14243 \text{ nm}}\right) = \boxed{151 \text{ eV}}$$

Part f)

$$\lambda_f = \lambda_i + \lambda_C(1 - \cos(\theta)) = (0.1400 \text{ nm}) + (0.00243 \text{ nm})(1 - \cos(180°)) = \boxed{0.14486 \text{ nm}}$$

$$K = hc\left(\frac{1}{\lambda_i} - \frac{1}{\lambda_f}\right) = (1240 \text{ eV} \cdot \text{nm})\left(\frac{1}{0.1400 \text{ nm}} - \frac{1}{0.14486 \text{ nm}}\right) = \boxed{297 \text{ eV}}$$

REFLECT

Since the wavelength of a photon is inversely proportional to its energy, the maximum wavelength of the scattered photon and the maximum kinetic energy of the scattered electrons both occur when the scattering angle is 180 degrees.

Get Help: Picture It – Compton Scattering

26.67

SET UP

An electron ($m_e = 9.11 \times 10^{-31}$ kg) has a speed of $v = 0.00730c$. Its de Broglie wavelength is equal to Planck's constant divided by its momentum.

SOLVE

$$\lambda = \frac{h}{p} = \frac{h}{m_e v} = \frac{6.63 \times 10^{-34} \text{ J} \cdot \text{s}}{(9.11 \times 10^{-31} \text{ kg})(0.00730)\left(3.00 \times 10^8 \frac{\text{m}}{\text{s}}\right)} = \boxed{3.32 \times 10^{-10} \text{ m}}$$

REFLECT

A distance of 3 angstroms (1 angstrom = 0.1 nm) is a little larger than a bond length in a molecule; this is why electrons are commonly used in microscopy to probe molecular structure. The exact de Broglie wavelength of the electron can be tuned by changing its speed.

Get Help: P'Cast 26.6 – A Slow Hummingbird

26.71

SET UP

We are the given the kinetic energies of six different electrons—1 eV, 10 eV, 100 eV, 1000 eV, 1×10^6 eV, and 1×10^9 eV—and asked to find the de Broglie wavelength, $\lambda = \dfrac{h}{p}$, of each one.

First, we need to determine whether or not we need to use the expression for the classical or relativistic momentum; relativistic effects should be included if the speed of the particle is about 10% the speed of light. If the speed we calculate from the kinetic energy is less than $0.1c$, then we can use the classical expressions for the kinetic energy $\left(K = \dfrac{1}{2}m_e v^2 \right)$ and momentum ($p = m_e v$). If the speed is larger than $0.1c$, we need to use the relativistic expressions for the kinetic energy, $K = (\gamma - 1)m_e c^2$, and momentum, $p = \gamma m_e v$.

SOLVE

Part a)

Speed:

$$K = \frac{1}{2}m_e v^2$$

$$v = \sqrt{\frac{2K}{m_e}} = \sqrt{\frac{2\left(1 \text{ eV} \times \dfrac{1.6 \times 10^{-19} \text{ J}}{1 \text{ eV}} \right)}{9.11 \times 10^{-31} \text{ kg}}} = 5.93 \times 10^5 \frac{\text{m}}{\text{s}}$$

Ratio of v to c:

$$\frac{v}{c} = \frac{\left(5.93 \times 10^5 \dfrac{\text{m}}{\text{s}} \right)}{\left(3.00 \times 10^8 \dfrac{\text{m}}{\text{s}} \right)} = 1.98 \times 10^{-3}, \text{ so we can safely ignore relativistic effects.}$$

de Broglie wavelength:

$$\lambda = \frac{h}{p} = \frac{h}{m_e v} = \frac{6.63 \times 10^{-34} \text{ J} \cdot \text{s}}{(9.11 \times 10^{-31} \text{ kg})\left(5.93 \times 10^5 \dfrac{\text{m}}{\text{s}} \right)} = \boxed{1.23 \times 10^{-9} \text{ m}}$$

Part b)

Speed:

$$K = \frac{1}{2}m_e v^2$$

$$v = \sqrt{\frac{2K}{m_e}} = \sqrt{\frac{2\left(10 \text{ eV} \times \dfrac{1.6 \times 10^{-19} \text{ J}}{1 \text{ eV}} \right)}{9.11 \times 10^{-31} \text{ kg}}} = 1.87 \times 10^6 \frac{\text{m}}{\text{s}}$$

Ratio of v to c:

$$\frac{v}{c} = \frac{\left(1.87 \times 10^6 \dfrac{\text{m}}{\text{s}} \right)}{\left(3.00 \times 10^8 \dfrac{\text{m}}{\text{s}} \right)} = 6.23 \times 10^{-3}, \text{ so we can safely ignore relativistic effects.}$$

de Broglie wavelength:

$$\lambda = \frac{h}{p} = \frac{h}{m_e v} = \frac{6.63 \times 10^{-34} \text{ J} \cdot \text{s}}{(9.11 \times 10^{-31} \text{ kg})\left(1.87 \times 10^6 \frac{\text{m}}{\text{s}}\right)} = \boxed{3.88 \times 10^{-10} \text{ m}}$$

Part c)

Speed:

$$K = \frac{1}{2} m_e v^2$$

$$v = \sqrt{\frac{2K}{m_e}} = \sqrt{\frac{2\left(100 \text{ eV} \times \frac{1.6 \times 10^{-19} \text{ J}}{1 \text{ eV}}\right)}{9.11 \times 10^{-31} \text{ kg}}} = 5.93 \times 10^6 \frac{\text{m}}{\text{s}}$$

Ratio of v to c:

$$\frac{v}{c} = \frac{\left(5.93 \times 10^6 \frac{\text{m}}{\text{s}}\right)}{\left(3.00 \times 10^8 \frac{\text{m}}{\text{s}}\right)} = 1.98 \times 10^{-2}, \text{ so we can safely ignore relativistic effects.}$$

de Broglie wavelength:

$$\lambda = \frac{h}{p} = \frac{h}{m_e v} = \frac{6.63 \times 10^{-34} \text{ J} \cdot \text{s}}{(9.11 \times 10^{-31} \text{ kg})\left(5.93 \times 10^6 \frac{\text{m}}{\text{s}}\right)} = \boxed{1.23 \times 10^{-10} \text{ m}}$$

Part d)

Speed:

$$K = \frac{1}{2} m_e v^2$$

$$v = \sqrt{\frac{2K}{m_e}} = \sqrt{\frac{2\left(1000 \text{ eV} \times \frac{1.6 \times 10^{-19} \text{ J}}{1 \text{ eV}}\right)}{9.11 \times 10^{-31} \text{ kg}}} = 1.87 \times 10^7 \frac{\text{m}}{\text{s}}$$

Ratio of v to c:

$$\frac{v}{c} = \frac{\left(1.87 \times 10^7 \frac{\text{m}}{\text{s}}\right)}{\left(3.00 \times 10^8 \frac{\text{m}}{\text{s}}\right)} = 6.23 \times 10^{-2}.$$

This speed is around 6% of the speed of light. For higher kinetic energies, we should start including relativistic effects in our calculations.

de Broglie wavelength:

$$\lambda = \frac{h}{p} = \frac{h}{m_e v} = \frac{6.63 \times 10^{-34} \text{ J} \cdot \text{s}}{(9.11 \times 10^{-31} \text{ kg})\left(1.87 \times 10^7 \frac{\text{m}}{\text{s}}\right)} = \boxed{3.88 \times 10^{-11} \text{ m}}$$

Part e)

Speed from the relativistic kinetic energy:

$$K = (\gamma - 1)m_e c^2 = \left(\frac{1}{\sqrt{1 - \left(\frac{v}{c}\right)^2}} - 1\right)m_e c^2$$

$$K + m_e c^2 = \frac{m_e c^2}{\sqrt{1 - \left(\frac{v}{c}\right)^2}}$$

$$\sqrt{1 - \left(\frac{v}{c}\right)^2} = \frac{m_e c^2}{K + m_e c^2}$$

$$1 - \left(\frac{v}{c}\right)^2 = \left(\frac{m_e c^2}{K + m_e c^2}\right)^2$$

$$v = c\sqrt{1 - \left(\frac{m_e c^2}{K + m_e c^2}\right)^2}$$

$$= \left(3.00 \times 10^8 \frac{\text{m}}{\text{s}}\right)\sqrt{1 - \left(\frac{(9.11 \times 10^{-31} \text{ kg})\left(3.00 \times 10^8 \frac{\text{m}}{\text{s}}\right)^2}{\left(1 \times 10^6 \text{ eV} \times \frac{1.6 \times 10^{-19} \text{ J}}{1 \text{ eV}}\right) + (9.11 \times 10^{-31} \text{ kg})\left(3.00 \times 10^8 \frac{\text{m}}{\text{s}}\right)^2}\right)^2}$$

$$= 2.82 \times 10^8 \frac{\text{m}}{\text{s}}$$

Ratio of v to c:

$$\frac{v}{c} = \frac{\left(2.82 \times 10^8 \frac{\text{m}}{\text{s}}\right)}{\left(3.00 \times 10^8 \frac{\text{m}}{\text{s}}\right)} = 0.94$$

Relativistic gamma:

$$\gamma = \frac{1}{\sqrt{1 - \left(\frac{v}{c}\right)^2}} = \frac{1}{\sqrt{1 - (0.94)^2}} = 2.93$$

de Broglie wavelength:

$$\lambda = \frac{h}{p} = \frac{h}{\gamma m_e v} = \frac{6.63 \times 10^{-34} \text{ J} \cdot \text{s}}{(2.93)(9.11 \times 10^{-31} \text{ kg})\left(2.82 \times 10^8 \frac{\text{m}}{\text{s}}\right)} = \boxed{8.80 \times 10^{-13} \text{ m}}$$

Part f)

Speed:

$$v = c\sqrt{1 - \left(\frac{m_e c^2}{K + m_e c^2}\right)^2}$$

$$= \left(3.00 \times 10^8 \frac{m}{s}\right)\sqrt{1 - \left(\frac{(9.11 \times 10^{-31}\ kg)\left(3.00 \times 10^8 \frac{m}{s}\right)^2}{\left(1 \times 10^9\ eV \times \frac{1.6 \times 10^{-19}\ J}{1\ eV}\right) + (9.11 \times 10^{-31}\ kg)\left(3.00 \times 10^8 \frac{m}{s}\right)^2}\right)^2}$$

$$= 2.9999996 \times 10^8 \frac{m}{s}$$

Ratio of v to c:

$$\frac{v}{c} = \frac{\left(2.9999996 \times 10^8 \frac{m}{s}\right)}{\left(3.00 \times 10^8 \frac{m}{s}\right)} = 0.99999987$$

Relativistic gamma:

$$\gamma = \frac{1}{\sqrt{1 - \left(\frac{v}{c}\right)^2}} = \frac{1}{\sqrt{1 - (0.99999987)^2}} = 1961$$

de Broglie wavelength:

$$\lambda = \frac{h}{p} = \frac{h}{\gamma m_e v} = \frac{6.63 \times 10^{-34}\ J \cdot s}{(1961)(9.11 \times 10^{-31}\ kg)\left(2.9999996 \times 10^8 \frac{m}{s}\right)} = \boxed{1.24 \times 10^{-14}\ m}$$

REFLECT

The de Broglie wavelength is inversely proportional to the speed of the particle; the kinetic energy is proportional to the speed of the particle. Therefore, as the kinetic energy increases, the de Broglie wavelength of the particle decreases.

Get Help: P'Cast 26.6 – A Slow Hummingbird

26.77

SET UP

The wavelength emitted by an electron flipping spin states in the ground state of hydrogen is $\lambda = 0.21$ m. The energy difference between energy states is equal to $\Delta E = \frac{hc}{\lambda}$.

SOLVE

$$\Delta E = \frac{hc}{\lambda} = \frac{(6.63 \times 10^{-34}\ J \cdot s)\left(3.00 \times 10^8 \frac{m}{s}\right)}{0.21\ m}$$

$$= 9.47 \times 10^{-21}\ J \times \frac{1\ eV}{1.6 \times 10^{-19}\ J} = \boxed{5.90 \times 10^{-6}\ eV}$$

REFLECT
Light with a wavelength of 21 cm is in the microwave region of the spectrum.

26.81

SET UP

The Balmer series results from transitions of electrons in hydrogen from higher energy states to the $n = 2$ level. The Rydberg formula, $\frac{1}{\lambda} = R_H\left(\frac{1}{n^2} - \frac{1}{m^2}\right)$, where $R_H = 1.09737 \times 10^7 \, \text{m}^{-1}$, gives the wavelengths of the photons associated with a transition from the mth to the nth energy level.

SOLVE

Rydberg formula:

$$\frac{1}{\lambda} = R_H\left(\frac{1}{n^2} - \frac{1}{m^2}\right)$$

From $m = 3$ to $n = 2$:

$$\lambda = \left[R_H\left(\frac{1}{n^2} - \frac{1}{m^2}\right)\right]^{-1} = \left[(1.09737 \times 10^7 \, \text{m}^{-1})\left(\frac{1}{(2)^2} - \frac{1}{(3)^2}\right)\right]^{-1}$$

$$= \boxed{6.56 \times 10^{-7} \, \text{m} = 656 \, \text{nm}}$$

From $m = 4$ to $n = 2$:

$$\lambda = \left[R_H\left(\frac{1}{n^2} - \frac{1}{m^2}\right)\right]^{-1} = \left[(1.09737 \times 10^7 \, \text{m}^{-1})\left(\frac{1}{(2)^2} - \frac{1}{(4)^2}\right)\right]^{-1}$$

$$= \boxed{4.86 \times 10^{-7} \, \text{m} = 486 \, \text{nm}}$$

From $m = 5$ to $n = 2$:

$$\lambda = \left[R_H\left(\frac{1}{n^2} - \frac{1}{m^2}\right)\right]^{-1} = \left[(1.09737 \times 10^7 \, \text{m}^{-1})\left(\frac{1}{(2)^2} - \frac{1}{(5)^2}\right)\right]^{-1}$$

$$= \boxed{4.34 \times 10^{-7} \, \text{m} = 434 \, \text{nm}}$$

From $m = 6$ to $n = 2$:

$$\lambda = \left[R_H\left(\frac{1}{n^2} - \frac{1}{m^2}\right)\right]^{-1} = \left[(1.09737 \times 10^7 \, \text{m}^{-1})\left(\frac{1}{(2)^2} - \frac{1}{(6)^2}\right)\right]^{-1}$$

$$= \boxed{4.10 \times 10^{-7} \, \text{m} = 410 \, \text{nm}}$$

REFLECT

The first four lines in the Balmer series are in the visible range, ranging from red to blue to violet; the remaining ones are in the ultraviolet region.

26.85

SET UP

The standard form of the Rydberg formula gives the wavelength of the emitted photons. Using $c = \lambda f$ we can express the Rydberg formula in terms of the frequency of the emitted photons.

SOLVE

$$\frac{1}{\lambda} = R_H\left(\frac{1}{n^2} - \frac{1}{m^2}\right)$$

$$f = \frac{c}{\lambda} = cR_H\left(\frac{1}{n^2} - \frac{1}{m^2}\right) = \left(2.9979 \times 10^8\frac{m}{s}\right)(1.09737 \times 10^7 \text{ m}^{-1})\left(\frac{1}{n^2} - \frac{1}{m^2}\right)$$

$$\boxed{f = (3.2898 \times 10^{15} \text{ Hz})\left(\frac{1}{n^2} - \frac{1}{m^2}\right)}$$

REFLECT

Comparing the frequency of photons is a bit more intuitive than comparing their wavelengths since the energy is directly proportional to the frequency, but inversely proportional to the wavelength.

26.89

SET UP

The angular momentum of an electron in the nth Bohr orbit is $L_n = n\hbar$, where $\hbar = \dfrac{h}{2\pi} = 1.055 \times 10^{-34}$ J · s. The radius of the nth Bohr orbit of hydrogen is given by $r_n = n^2 a_0$, where $a_0 = 0.529 \times 10^{-10}$ m. The speed of an electron can be found by dividing its angular momentum by its mass and the radius of its orbit; therefore, the speed of an electron in the nth Bohr orbit of hydrogen is $v_n = \dfrac{L_n}{m_e r_n}$. To find these values for an electron in the tenth Bohr orbit, we should plug in $n = 10$ and solve.

SOLVE

Angular momentum of an electron in the tenth Bohr orbit:

$$L_n = n\hbar$$

$$L_{10} = 10\hbar = 10(1.055 \times 10^{-34} \text{ J} \cdot \text{s}) = \boxed{1.055 \times 10^{-33} \text{ J} \cdot \text{s}}$$

Radius of the tenth Bohr orbit of hydrogen:

$$r_n = n^2 a_0$$

$$r_{10} = (10)^2(0.529 \times 10^{-10} \text{ m}) = 5.29 \times 10^{-9} \text{ m}$$

Speed of an electron in the tenth Bohr orbit:

$$v_n = \frac{L_n}{m_e r_n}$$

$$v_{10} = \frac{L_{10}}{m_e r_{10}} = \frac{1.066 \times 10^{-33}\,\text{J} \cdot \text{s}}{(9.11 \times 10^{-31}\,\text{kg})(5.29 \times 10^{-9}\,\text{m})} = \boxed{2.19 \times 10^5 \frac{\text{m}}{\text{s}}}$$

REFLECT

The general expression for the quantized radii of electron orbits is $r_n = \frac{n^2 a_0}{Z}$, where Z is the atomic number; for hydrogen, $Z = 1$.

26.93

SET UP

A hydrogen atom has an electron in the $n = 2$ state when it absorbs a photon, which promotes the electron to the $n = 4$ state. This electron will then emit light and relax back down to the $n = 1$ state. We can use the Rydberg equation to calculate the wavelength of the absorbed photon, as well as the wavelengths of the possible emitted photons. Since the Rydberg equation is used to calculate the wavelength of the *emitted* photon (*i.e.*, $n > m$), we need to either take the absolute value or reverse the terms to correctly calculate the wavelength of the *absorbed* photon.

SOLVE

Part a)

$$\frac{1}{\lambda} = R_H\left(\frac{1}{n^2} - \frac{1}{m^2}\right)$$

$$\lambda = \left[R_H\left(\frac{1}{n^2} - \frac{1}{m^2}\right)\right]^{-1} = \left|\left[(1.09737 \times 10^7\,\text{m}^{-1})\left(\frac{1}{(4)^2} - \frac{1}{(2)^2}\right)\right]^{-1}\right| =$$

$$= \boxed{4.86 \times 10^{-7}\,\text{m} = 486\,\text{nm}}$$

Part b)

Emission process #1: $n = 4 \rightarrow n = 3 \rightarrow n = 2 \rightarrow n = 1$

$$\lambda_{4\rightarrow3} = \left[(1.09737 \times 10^7\,\text{m}^{-1})\left(\frac{1}{3^2} - \frac{1}{4^2}\right)\right]^{-1} = 1.875 \times 10^{-6}\,\text{m} = \boxed{1875\,\text{nm}}$$

$$\lambda_{3\rightarrow2} = \left[(1.09737 \times 10^7\,\text{m}^{-1})\left(\frac{1}{2^2} - \frac{1}{3^2}\right)\right]^{-1} = 6.56 \times 10^{-7}\,\text{m} = \boxed{656\,\text{nm}}$$

$$\lambda_{2\rightarrow1} = \left[(1.09737 \times 10^7\,\text{m}^{-1})\left(\frac{1}{1^2} - \frac{1}{2^2}\right)\right]^{-1} = 1.22 \times 10^{-7}\,\text{m} = \boxed{122\,\text{nm}}$$

Emission process #2: $n = 4 \rightarrow n = 2 \rightarrow n = 1$

$$\lambda_{4\rightarrow2} = \left[(1.09737 \times 10^7\,\text{m}^{-1})\left(\frac{1}{2^2} - \frac{1}{4^2}\right)\right]^{-1} = 4.86 \times 10^{-6}\,\text{m} = \boxed{486\,\text{nm}}$$

$$\lambda_{2\rightarrow1} = 122\,\text{nm}$$

Emission process #3: $n = 4 \rightarrow n = 3 \rightarrow n = 1$

$$\lambda_{4 \rightarrow 3} = 1875 \text{ nm}$$

$$\lambda_{3 \rightarrow 1} = \left[(1.09737 \times 10^7 \text{ m}^{-1})\left(\frac{1}{1^2} - \frac{1}{3^2}\right)\right]^{-1} = 1.03 \times 10^{-7} \text{ m} = \boxed{103 \text{ nm}}$$

Emission process #4: $n = 4 \rightarrow n = 1$

$$\lambda_{4 \rightarrow 2} = \left[(1.09737 \times 10^7 \text{ m}^{-1})\left(\frac{1}{1^2} - \frac{1}{4^2}\right)\right]^{-1} = 9.72 \times 10^{-8} \text{ m} = \boxed{97.2 \text{ nm}}$$

REFLECT

We don't know ahead of time which relaxation process the electron will undergo on its way back to the $n = 1$ state; all are possible, but some are more probable than others.

26.103

SET UP

The speed of an electron in hydrogen traveling in a circular orbit of radius r_n is equal to $v_n = \frac{2\pi r_n}{T_n}$. From Equations 26-26 and 26-33, the radius and speed are also equal to $r_n = \frac{n^2 \hbar^2}{ke^2 m_e}$ and $v_n = \frac{ke^2}{n\hbar}$. Combining these with the fact that $f_n = \frac{1}{T_n}$ will allow us to derive an expression for the frequency of an electron revolving in the nth orbit of a hydrogenic atom.

SOLVE

$$v_n = \frac{2\pi r_n}{T_n}$$

$$f_n = \frac{1}{T_n} = \frac{v_n}{2\pi r_n} = \left(\frac{ke^2}{n\hbar}\right)\left(\frac{1}{2\pi}\right)\left(\frac{ke^2 m_e}{n^2\hbar^2}\right) = \boxed{\frac{k^2 e^4 m_e}{2\pi n^3 \hbar^3}}$$

REFLECT

We can also represent the frequency in terms of the Bohr radius: $f_n = \dfrac{ke^2}{2\pi n^3 \hbar a_0}$

26.109

SET UP

A gamma ray is converted into an electron and a positron, each of which have a mass $m_e = 9.11 \times 10^{-31}$ kg. The maximum wavelength of a gamma ray is the same as asking, "What's the least energetic gamma ray that can produce an electron and positron, each at rest?" The energy of the gamma ray is completely converted into the mass of the electron and the positron. We can set the sum of the rest energies of the electron and the positron equal to the energy of the gamma ray in order to solve for the maximum possible wavelength.

SOLVE

Energy of the gamma ray:

$$E = \frac{hc}{\lambda}$$

Minimum energy to create an electron and a positron:

$$E = m_e c^2 + m_p c^2 = m_e c^2 + m_e c^2 = 2m_e c^2$$

Minimum wavelength for the gamma ray:

$$\frac{hc}{\lambda} = 2m_e c^2$$

$$\lambda = \frac{h}{2m_e c} = \frac{6.63 \times 10^{-34} \text{ J} \cdot \text{s}}{2(9.11 \times 10^{-31} \text{ kg})\left(3.00 \times 10^8 \frac{\text{m}}{\text{s}}\right)} = \boxed{1.21 \times 10^{-12} \text{ m} = 1.21 \text{ pm}}$$

REFLECT

In the electromagnetic spectrum, gamma rays have a wavelength less than 10 pm, so our result is reasonable.

Get Help: P'Cast 26.9 – A Mercury Line

Chapter 27
Nuclear Physics

Conceptual Questions

27.9 Certainly the central ideas of binding energy and "Q values" of nuclear reactions would not have been possible to develop without the $E = mc^2$ concept. The Manhattan project in World War II, where nuclear weapons were designed and successfully tested, would not have been possible had it not been for Einstein's theory of the equivalence of mass and energy (as well as his letter to President Roosevelt endorsing this project in the early 1940s).

27.13 Consider the typical β decay of a neutron: $n \rightarrow p + \beta^- + \bar{\nu}_e$. If the electron ($\beta^-$) were the only decay product, application of conservation of energy and momentum to the two-body decay would require that the β^- particle be ejected with a single unique energy. Instead, we observe experimentally that β^- particles are produced with energies that range from zero to a maximum value. Further, because the original neutron had spin 1/2, conservation of angular momentum would be violated if the final decay products consisted of only the two particles p and β^-, each with spin 1/2.

27.17 Because the atomic masses include Z electron masses in their values, it is already taken into account. There is 1 electron mass included in the atomic mass of the proton. There is no need to add in another one to account for the beta particle.

Multiple-Choice Questions

27.21 D (Both fusion and fission release energy). Energy is released in either process in order to form more stable products.

27.25 B (less than). Energy is released in a spontaneous fusion process. Mass and energy are equivalent through $E = mc^2$.

Estimation/Numerical Questions

27.31 An atomic nucleus is about 5×10^{17} times more dense than an atom.

27.37

$$\frac{N(t)}{N_0} = \left[\frac{1}{2}\right]^{\frac{t}{\tau_{1/2}}}$$

$$t = \tau_{1/2} \frac{\ln\left(\dfrac{N(t)}{N_0}\right)}{\ln\left(\dfrac{1}{2}\right)}$$

Part a)

$$t = (1 \text{ day})\frac{\ln\left(\frac{1}{2}\right)}{\ln(0.625)} = 0.678 \text{ days}$$

Part b)

$$t = (1 \text{ day})\frac{\ln(0.0625)}{\ln\left(\frac{1}{2}\right)} = 4 \text{ days}$$

Problems

27.43

SET UP

A nucleus is approximately spherical with a radius $r = r_0 A^{\frac{1}{3}}$, where $r_0 = 1.2 \times 10^{-15}$ m and A is the mass number. The mass density of the nucleus is equal to its mass divided by its volume. We can approximate the total mass of the nucleus as A multiplied by the average mass of a nucleon. The mass of a proton is $m_p = 1.6726 \times 10^{-27}$ kg, and the mass of a neutron is $m_n = 1.6749 \times 10^{-27}$ kg.

SOLVE

$$\rho = \frac{m}{V} = \frac{Am_{avg}}{\left(\frac{4}{3}\pi r^3\right)} = \frac{3A\left(\dfrac{m_p + m_n}{2}\right)}{4\pi(r_0 A^{\frac{1}{3}})^3} = \frac{3(m_p + m_n)}{8\pi r_0^3}$$

$$= \frac{3((1.6726 \times 10^{-27} \text{ kg}) + (1.6749 \times 10^{-27} \text{ kg}))}{8\pi(1.2 \times 10^{-15} \text{ m})^3} = \boxed{2.3 \times 10^{17}\frac{\text{kg}}{\text{m}^3}}$$

$$2.3 \times 10^{17}\frac{\text{kg}}{\text{m}^3} \times \frac{2.2 \text{ lb}}{1 \text{ kg}} \times \frac{1 \text{ ton}}{2000 \text{ lb}} \times \left(\frac{1 \text{ m}}{100 \text{ cm}}\right)^3 \times \left(\frac{2.54 \text{ cm}}{1 \text{ in}}\right)^3 = \boxed{4.2 \times 10^9\frac{\text{tons}}{\text{in}^3}}$$

REFLECT

The nucleus is extremely dense, so an answer on the order of 10^{17} is reasonable. Note that the density did not depend upon the mass number.

Get Help: P'Cast 27.2 – Nuclear Density

27.47

SET UP

Carbon-12 has a mass of $m_{atom} = 12.000000$ u and is made up of 6 neutrons, 6 protons, and 6 electrons. The binding energy of the atom is the difference in energy between the component parts of the atom and the atom itself, $E_B = (Nm_n + Zm_{^1H} - m_{atom})c^2$, where $m_n = 1.008665$ u and $m_{^1H} = 1.007825$ u. The conversion between u and MeV is 1 u = 931.494 MeV/c^2.

SOLVE

$$E_B = (Nm_n + Zm_{^1H} - m_{atom})c^2$$

$$E_B = (6(1.008665 \text{ u}) + 6(1.007825 \text{ u}) - (12.000000 \text{ u}))c^2$$

$$= (0.09894 \text{ u})c^2 \times \frac{\left(931.494\frac{\text{MeV}}{c^2}\right)}{1 \text{ u}} = \boxed{92.2 \text{ MeV}}$$

REFLECT

The larger the binding energy is, the more stable the nucleus.

27.55

SET UP

Thorium-232 undergoes the following nuclear fission reaction: $^{232}\text{Th} + \text{n} \rightarrow {}^{99}\text{Kr} + {}^{124}\text{Xe} + \_\_$. We can determine the missing product by calculating its expected atomic number Z and mass number A. The sum of the atomic numbers of the products must equal the sum of the atomic numbers of the reactants; the same must be true for the mass numbers. The atomic numbers of thorium, krypton, and xenon are 90, 36, and 54, respectively. The mass numbers of each species are to the top left of the atomic symbol; a neutron has a mass number of 1. The energy Q released by the reaction is equal to the total energy of the reactants minus the total energy of the products. The atomic masses are thorium-232 = 232.038051 u, a neutron = 1.008665 u, krypton-99 = 98.957606 u, and xenon-124 = 123.905894 u.

SOLVE

Atomic number:

$$Z = 90 - (36 + 54) = 0$$

Since $Z = 0$, the missing product will be a neutron.

Mass number:

$$A = 232 + 1 - (99 + 124) = 10$$

The total mass number of the missing must be 10. We already know it is a neutron, which has a mass number of 1, so the missing product must be 10 neutrons, or $\boxed{10n}$.

Energy released:

$$Q = [(232.038051 \text{ u}) + (1.008665 \text{ u}) - ((98.957606 \text{ u}) + (123.905894 \text{ u}) + 10(1.008665 \text{ u}))]c^2$$

$$= (0.096566 \text{ u})c^2 \times \frac{\left(931.494\frac{\text{MeV}}{c^2}\right)}{1 \text{ u}} = \boxed{89.95 \text{ MeV}}$$

REFLECT

The total number of protons and the total number of neutrons must remain constant.

Get Help: P'Cast 27.5 – Fission to Xenon and Beyond

27.61

SET UP

The fission of a uranium-235 releases 185×10^6 eV of energy. In order to determine how many kilograms of uranium-235 are necessary to produce 1000×10^6 W of power continuously for one year, we must first determine the total amount of energy required by multiplying the power by the time interval. Each fission reaction uses 1 nucleus; we can use Avogadro's number and the molar mass of uranium-235 in order to calculate the mass.

SOLVE

Total energy required:

$$E = P\Delta t = (1000 \times 10^6 \text{ W})\left(1 \text{ yr} \times \frac{365.25 \text{ days}}{1 \text{ yr}} \times \frac{24 \text{ hr}}{1 \text{ day}} \times \frac{3600 \text{ s}}{1 \text{ hr}}\right) = 3.1558 \times 10^{16} \text{ J}$$

Total number of fission reactions necessary:

$$\frac{3.1558 \times 10^{16} \text{ J}}{185 \times 10^6 \text{ eV}} \times \frac{1 \text{ eV}}{1.6 \times 10^{-19} \text{ J}} = 1.066 \times 10^{27} \text{ reactions}$$

Mass of uranium-235 required:

$$1.066 \times 10^{27} \text{ reactions} \times \frac{1 \text{ nucleus U} - 235}{1 \text{ reaction}} \times \frac{235 \text{ g}}{6.02 \times 10^{23} \text{ nuclei}} = \boxed{4.16 \times 10^5 \text{ g} = 416 \text{ kg}}$$

REFLECT

We've assumed that the fission reaction is 100% efficient, which means our answer is the *minimum* mass required.

27.67

SET UP

The deuterium-tritium (D-T) fusion reaction forms helium-4 and a neutron from deuterium and tritium: $D + T \rightarrow {}^4\text{He} + n$; the reaction releases about 20×10^6 eV of energy. We can use this information, along with Avogadro's number and the molar mass of tritium (1 mol = 3 g) in order to calculate the amount of tritium necessary to create 10^{14} J of energy.

SOLVE

$$10^{14} \text{ J} \times \frac{1 \text{ eV}}{1.6 \times 10^{-19} \text{ J}} \times \frac{1 \text{ nucleus T}}{20 \times 10^6 \text{ eV}} \times \frac{3 \text{ g}}{6.02 \times 10^{23} \text{ nuclei T}} = \boxed{156 \text{ g}}$$

REFLECT

This reaction is the simplest fusion reaction to perform on Earth and, because of this, has great promise for potential future sources of nuclear energy.

27.73

SET UP

A certain radioisotope has a decay constant of $\lambda = 0.00334 \text{ s}^{-1}$. The half-life is equal to $\tau_{1/2} = \frac{\ln(2)}{\lambda}$.

SOLVE

$$\tau_{1/2} = \frac{\ln(2)}{\lambda} = \frac{\ln(2)}{0.00334 \text{ s}^{-1}} = \boxed{208 \text{ s}} \times \frac{1 \text{ hr}}{3600 \text{ s}} \times \frac{1 \text{ day}}{24 \text{ hr}} = \boxed{0.00240 \text{ days}}$$

REFLECT
This is about 3.5 min.

27.77

SET UP

A patient is injected with 7.88 μCi of iodine-131 that has a half-life of $\tau_{1/2} = 8.02$ days. We want to calculate the expected decay rate in the patient's thyroid after 30 days. The decay rate exponentially decays with time, $R = R_0 e^{-\lambda t}$, where R_0 is the initial rate and $\lambda = \dfrac{\ln(2)}{\tau_{1/2}}$. Only 90% of the initial dose makes its way to the thyroid, which means $R_0 = (0.90)(7.88 \ \mu\text{Ci})$.

SOLVE
Decay constant:

$$\lambda = \frac{\ln(2)}{\tau_{1/2}} = \frac{\ln(2)}{8.02 \text{ days}} = 0.0864 \text{ days}^{-1}$$

Decay rate after 30 days:

$$R = R_0 e^{-\lambda t} = (0.90)(7.88 \ \mu\text{Ci}) e^{-(0.0864 \text{ days}^{-1})(30 \text{ days})} = \boxed{0.531 \ \mu\text{Ci}}$$

REFLECT
The number of decays per second depends upon the number of atoms present; if the number of atoms decreases exponentially with time, so too should the decay rate.

27.81

SET UP

A sample of radon-222 has a decay rate of $R = 485$ counts/min. The decay rate of the sample is equal to the product of the decay constant λ and the number of nuclei N. The half-life of radon-222 from Appendix C is $\tau_{1/2} = 3.823$ days. After finding λ from the half-life, we can divide the decay rate by it in order to calculate the number of nuclei in the sample.

SOLVE
Decay constant:

$$\lambda = \frac{\ln(2)}{\tau_{1/2}} = \frac{\ln(2)}{\left(3.823 \text{ days} \times \dfrac{24 \text{ hr}}{1 \text{ day}} \times \dfrac{3600 \text{ s}}{1 \text{ hr}} \right)} = 2.0985 \times 10^{-6} \text{ s}^{-1}$$

Number of nuclei:

$$R = \lambda N$$

$$N = \frac{R}{\lambda} = \frac{\left(485 \dfrac{\text{counts}}{\text{min}} \times \dfrac{1 \text{ min}}{60 \text{ s}} \right)}{2.0985 \times 10^{-6} \text{ s}^{-1}} = \boxed{3.85 \times 10^6 \text{ nuclei}}$$

REFLECT

The units of "counts/second" are equivalent to becquerels, as long as the detector detects 100% of the radioactive decays.

27.87

SET UP

The approximate radius of a uranium-238 nucleus is given by $r = r_0 A^{\frac{1}{3}}$, where $r_0 = 1.2$ fm and A is the mass number. The magnitude of the force between two protons located at opposite ends of the nucleus can be calculated from Coulomb's law. If this were the only force acting on the protons, we can calculate their resulting acceleration from Newton's second law. Finally, the nucleus is held together by the strong nuclear force, which allows the nucleons to fuse together to form nuclei.

SOLVE

Part a)

$$r = r_0 A^{\frac{1}{3}} = (1.2 \text{ fm})(238)^{\frac{1}{3}} = \boxed{7.4 \text{ fm}}$$

Part b)

$$F = \frac{k(e)(e)}{r^2} = \frac{\left(9.0 \times 10^9 \frac{\text{N} \cdot \text{m}^2}{\text{C}^2}\right)(1.6 \times 10^{-19} \text{ C})^2}{(2(7.4 \times 10^{-15} \text{ m}))^2} = \boxed{1.0 \text{ N}}$$

Part c)

$$\sum F = F = m_p a$$

$$a = \frac{F}{m_p} = \frac{1.0 \text{ N}}{(1.67 \times 10^{-27} \text{ kg})} = \boxed{6.2 \times 10^{26} \frac{\text{m}}{\text{s}^2}}$$

Part d) The strong nuclear force holds them together.

REFLECT

The weight of a proton is about 10^{-26} N, which means the electrostatic force between the two farthest protons in the nucleus is 26 orders of magnitude larger! For a comparison, the mass of the Earth is "only" 23 orders of magnitude larger than the mass of a person.

Get Help: P'Cast 27.1 – Nuclear Radii

27.91

SET UP

An isotope of element 117 was created that had 176 neutrons and a half-life of $\tau_{1/2} = 14 \times 10^{-3}$ s. The mass number A of the isotope is the sum of the atomic number and the number of neutrons. The approximate radius of the nucleus is given by $r = r_0 A^{\frac{1}{3}}$, where $r_0 = 1.2$ fm. The fraction of the sample that remains as a function of time is related to the half-life by $\frac{N(t)}{N_0} = \left[\frac{1}{2}\right]^{\frac{t}{\tau_{1/2}}}$; we can use this expression to calculate the percent of the created isotope left after 1.0 s.

SOLVE

Part a)

Mass number:

$$A = N + Z = 176 + 117 = 293$$

Radius of the nucleus:

$$r = r_0 A^{\frac{1}{3}} = (1.2 \text{ fm})(293)^{\frac{1}{3}} = \boxed{8.0 \text{ fm}}$$

Part b)

$$\frac{N(1.0 \text{ s})}{N_0} = \left[\frac{1}{2}\right]^{\frac{1.0 \text{ s}}{14 \times 10^{-3} \text{ s}}} = 3.1 \times 10^{-22} = \boxed{3.1 \times 10^{-20} \text{ \%}}$$

REFLECT

A time interval of 1.0 s is a little over 71 half-lifes of the new element, so we should expect the amount of the element created to be extremely small.

27.95

SET UP

Radiocarbon dating was used to determine the age of a bone sample found in a cave. The results showed that the level of carbon-14 present was 2.35% of its present-day level. The fraction of the sample that remains as a function of time is related to the half-life by $\frac{N(t)}{N_0} = \left[\frac{1}{2}\right]^{\frac{t}{\tau_{1/2}}}$; we can use this expression to calculate the age of the bones. The half-life of carbon-14 from Appendix C is $\tau_{1/2} = 5730$ yr.

SOLVE

$$\frac{N(t)}{N_0} = \left[\frac{1}{2}\right]^{\frac{t}{\tau_{1/2}}} = 2^{-\frac{t}{\tau_{1/2}}}$$

$$\ln\left(\frac{N(t)}{N_0}\right) = -\frac{t}{\tau_{1/2}}\ln(2)$$

$$t = -\tau_{1/2}\frac{\ln\left(\frac{N(t)}{N_0}\right)}{\ln(2)} = -(5730 \text{ yr})\frac{\ln(0.0235)}{\ln(2)} = \boxed{31,000 \text{ yr}}$$

REFLECT

Radiocarbon dating is commonly used to date organic samples. Plants exchange carbon with the atmosphere through photosynthesis and will, therefore, contain the various isotopes of carbon in the same ratio as the atmosphere. As long as the plant is alive, the ratio should remain relatively constant. When the plant dies, photosynthesis no longer takes place, and the amount of carbon-14 will decrease with time because it is radioactive. Some animals eat plants, and other animals will eat these animals, which is how carbon-14 incorporates itself into animals.

27.103

SET UP

We are told that electron capture by a proton is not allowed in nature, which means the reaction $e^- + p \rightarrow n + \nu$, where ν represents a particle of negligible mass called the neutrino, does not occur. We can calculate the energy Q released by this process in order to better understand why this process does not take place. The masses of a neutron, proton, and electron in amu are $m_n = 1.008665$ u, $m_p = 1.007277$ u, and $m_e = 0.005486$ u.

SOLVE

$$Q = [(0.0005486 \text{ u}) + (1.007277 \text{ u}) - (1.008665 \text{ u})]c^2$$

$$= (-0.0008934 \text{ u})c^2 \times \frac{\left(931.494 \, \frac{\text{MeV}}{c^2}\right)}{1 \text{ u}} = -0.7819 \text{ MeV}$$

Since the $\boxed{\text{energy released we calculated is negative, this reaction will not proceed}}$.

REFLECT

If this reaction were to occur, all low energy electrons could react with protons and turn them into neutrons. This would destroy chemistry as we know it!

Chapter 28
Particle Physics

Conceptual Questions

28.5 There are 10 combinations: uud, udd, ddd, uuu, uus, dds, uds, uss, dss, and sss.

Multiple-Choice Questions

28.11 B (meson). A particle composed of two quarks is classified as a meson.

28.15 A (strong). Quarks are attracted to each other through the strong force.

Estimation/Numerical Questions

28.21 The ratios of the meson masses are K:π = 3.6, D:K = 3.8, B:D = 2.8. The ratio of the quark masses, from Table 28-1, are s:d = 20, c:s = 13, b:c = 3.8. No, the ratios of the meson masses do not follow the ratio of the quark masses. The mass of the quarks alone does not account for the mass of the mesons they comprise.

Problems

28.25

SET UP

We can determine whether or not the reaction p $\rightarrow$ e$^+$ + γ is possible by seeing if the charge, baryon number, and the lepton number are conserved. Every matter baryon has a baryon number of $+1$, and particles that are not baryons have a baryon number of 0. A positron has an electron-lepton number of -1, and a particle that is not a lepton has an electron-lepton number of 0.

SOLVE

Charge: A proton and a positron each have a charge of $+e$. A photon is uncharged. Charge is conserved.

Baryon number: A proton has a baryon number of $+1$. A positron and a photon each have a baryon number of 0. The baryon number is not conserved.

Electron-Lepton number: A proton and a photon each have an electron-lepton number of 0. A positron has an electron-lepton number of -1. The electron-lepton number is not conserved.

No, the reaction is $\boxed{\text{not possible}}$ because neither the baryon number nor the electron-lepton number is conserved.

REFLECT

The sum of the baryon numbers and lepton numbers of the particles that participate in the process must equal the sum of the values of the particles present at the end of the process.

28.29

SET UP

We are asked to determine the quark structure of a K^0 meson; a meson must be composed of two quarks. We are told that the K meson is the lowest mass strange particle. Therefore, it must contain an s or an s-bar quark. A K^0 meson will be made up of quark-antiquark pair because it is neutrally charged. The lowest mass quarks are u (around 2 MeV/c^2) and d (around 5 MeV/c^2). Both an s quark and a d quark have a charge of $-1/3e$, whereas a u quark has a charge of $+2/3e$. In order to satisfy charge neutrality, the K^0 meson must contain either a d or d-bar quark. It turns out that the K^0 meson is made up of a d quark and an s-bar quark.

SOLVE

The quark structure of a K^0 meson is $\boxed{d\bar{s}}$.

REFLECT

The antiparticle of the K^0 meson is the $\overline{K}^0$ meson and is composed of a d-bar quark and an s quark, $\bar{d}s$.

28.33

SET UP

We can determine whether or not the reactions are possible by seeing if the charge, baryon number, and the lepton number are conserved. Every matter baryon has a baryon number of $+1$, and particles that are not baryons have a baryon number of 0. Matter leptons have a lepton number of $+1$, while antimatter leptons have a lepton number of -1. Particles that are not leptons have a lepton number of 0.

SOLVE

Part a)

$$n \rightarrow p + e^- + \bar{\nu}_e$$

Charge: A neutron and an antimatter neutrino each have a charge of 0. A proton has a charge of $+e$, and an electron has a charge of $-e$. Charge is conserved.

Baryon number: A neutron and proton each have a baryon number of $+1$. An electron and an antimatter electron neutrino each have a baryon number of 0. The baryon number is conserved.

Electron-Lepton number: A neutron and a proton each have an electron-lepton number of 0. An electron has an electron-lepton number of $+1$. An antimatter electron neutrino has an electron-lepton number of -1. The electron-lepton number is conserved.

This reaction is $\boxed{\text{possible}}$.

Part b)

$$\mu^- \rightarrow e^- + \bar{\nu}_e + \nu_\mu$$

Charge: A neutrino and an antimatter neutrino each have a charge of 0. A muon and an electron each have a charge of $-e$. Charge is conserved.

Baryon number: All of the particles involved have a baryon number of 0. The baryon number is conserved.

Lepton number: An electron has an electron-lepton number of $+1$. An antimatter electron neutrino has an electron-lepton number of -1. The electron-lepton number is conserved. A muon and a muon neutrino each have a muon-lepton number of $+1$. The muon-lepton number is conserved.

This reaction is $\boxed{\text{possible}}$.

Part c)

$$\pi^- \rightarrow \mu^- + \bar{\nu}_\mu$$

Charge: A negatively charged pion and a muon each have a charge of $-e$. An antimatter neutrino has a charge of 0. Charge is conserved.

Baryon number: All of the particles involved have a baryon number of 0. The baryon number is conserved.

Lepton number: A pion has a lepton number of 0. A muon has a muon-lepton number of $+1$. An antimatter muon neutrino has a muon-lepton number of -1. The muon-lepton number is conserved.

This reaction is $\boxed{\text{possible}}$.

REFLECT

All three reactions conserve charge, baryon number, and lepton number.

28.39

SET UP

A high-energy photon in the vicinity of a nucleus can create an electron-positron pair by pair production: $\gamma \rightarrow e^- + e^+$. The minimum energy required for this process is equal to the rest energy of an electron and a positron. The mass of an electron and the mass of a positron are both equal to $m_e = 9.11 \times 10^{-31}$ kg. Since the electron and positron are created at rest, the nucleus is necessary to conserve momentum.

SOLVE

Part a)

$$E = m_e c^2 + m_e c^2 = 2m_e c^2 = 2(9.11 \times 10^{-31} \text{ kg})\left(3.00 \times 10^8 \frac{\text{m}}{\text{s}}\right)^2$$

$$= \boxed{1.64 \times 10^{-13} \text{ J}} \times \frac{1 \text{ eV}}{1.6 \times 10^{-19} \text{ J}} = 1.02 \times 10^6 \text{ eV} = \boxed{1.02 \text{ MeV}}$$

Part b) The nucleus is required to absorb the momentum so that conservation of momentum is not violated.

REFLECT

The excess energy of a photon with an energy higher than 1.02 MeV undergoing pair production goes into the kinetic energy of the electron and the positron.

28.43

SET UP

A proton-antiproton annihilation takes place and the resulting photons have a total energy of 2.5×10^9 eV. The energy of the photons is equal to the sum of the rest energies and the kinetic energies of the proton and antiproton. The mass of a proton and the mass of an antiproton are both equal to $m_p = 1.67 \times 10^{-27}$ kg. Using energy conservation, we can calculate the kinetic energy of the proton K_p and the kinetic energy of the antiproton $K_{\bar{p}}$ if they have the same kinetic energy or if $K_p = 1.25K_{\bar{p}}$.

SOLVE

Mass of proton in MeV/c^2:

$$E = m_p c^2 = (1.67 \times 10^{-27} \text{ kg})\left(3.00 \times 10^8 \frac{\text{m}}{\text{s}^2}\right)$$

$$= 1.503 \times 10^{-10} \text{ J} \times \frac{1 \text{ eV}}{1.6 \times 10^{-19} \text{ J}} = 9.38 \times 10^8 \text{ eV}$$

$$m_p = 9.38 \times 10^8 \frac{\text{eV}}{c^2}$$

Part a)

$$E = m_p c^2 + m_{\bar{p}} c^2 + K_p + K_{\bar{p}} = 2m_p c^2 + 2K_p$$

$$K_p = \frac{E - 2m_p c^2}{2} = \frac{(2.5 \times 10^9 \text{ eV}) - 2\left(9.38 \times 10^8 \frac{\text{eV}}{c^2}\right)c^2}{2} = \boxed{3.12 \times 10^8 \text{ eV} = 0.312 \text{ GeV}}$$

$$\boxed{K_p = K_{\bar{p}} = 0.312 \text{ GeV}}$$

Part b)

$$E = m_p c^2 + m_{\bar{p}} c^2 + K_p + K_{\bar{p}} = 2m_p c^2 + 1.25K_{\bar{p}} + K_{\bar{p}}$$

$$K_{\bar{p}} = \frac{E - 2m_p c^2}{2.25} = \frac{(2.5 \times 10^9 \text{ eV}) - 2\left(9.38 \times 10^8 \frac{\text{eV}}{c^2}\right)c^2}{2.25} = \boxed{2.77 \times 10^8 \text{ eV} = 0.277 \text{ GeV}}$$

$$\boxed{K_p = 1.25K_{\bar{p}} = 1.25(0.277 \text{ GeV}) = 0.347 \text{ GeV}}$$

REFLECT

Since the proton and antiproton annihilate one another, both their kinetic energy *and* rest energy are converted into photons.

28.47

SET UP

An electron and a positron undergo pair annihilation. The Feynman diagram should have two straight lines on the left representing the electron and positron that meet at a vertex. A wavy line representing the photon should then leave the vertex headed towards the right.

SOLVE

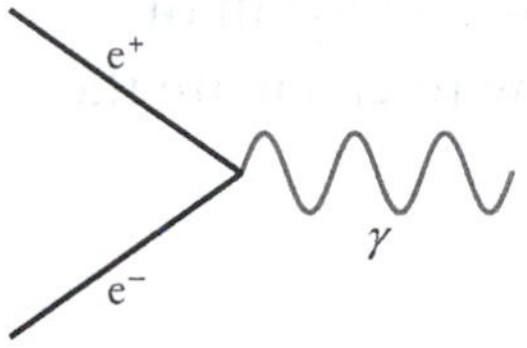

Figure 28-1 Problem 47

REFLECT

Time runs from left to right in a Feynman diagram, so the process starts with two particles and ends with a photon, as expected for an annihilation process.

28.51

SET UP

Sodium-22 decays by emitting a positron. In order to write out the decay, we need to apply the conservation of baryon number and electron-lepton number. The resulting nucleus needs to have a mass number of 22 in order to conserve baryon number, but the atomic number of the nucleus should decrease by 1 due to the emitted positron. Sodium has an atomic number of 11, which means the product nucleus should have an atomic number 10, corresponding to neon. Finally, the electron-lepton number of the positron is -1, so we need to also produce an electron neutrino in order to conserve lepton number. Initially, there are 5×10^{23} sodium-22 nuclei. The number of positrons emitted per second is equal to the number of sodium-22 decays per second, which is the activity of the sample. The decay constant is related to the half-life of sodium-22, $\tau_{1/2} = 2.60$ yr.

SOLVE

Decay:

$$^{22}\text{Na} \rightarrow \text{e}^+ + {}^{22}\text{Ne} + \nu_e$$

Emitted positrons per second:

$$\text{activity} = \lambda N = \left(\frac{\ln(2)}{\tau_{1/2}}\right)N = \frac{\ln(2)}{\left(2.60 \text{ yr} \times \dfrac{365.25 \text{ days}}{1 \text{ yr}} \times \dfrac{24 \text{ hr}}{1 \text{ day}} \times \dfrac{3600 \text{ s}}{1 \text{ hr}}\right)}(5 \times 10^{23}\ {}^{22}\text{Na})$$

$$= 4.22 \times 10^{15}\frac{{}^{22}\text{Na}}{\text{s}} \times \frac{1 \text{ e}^+}{1\ {}^{22}\text{Na}} = \boxed{4.22 \times 10^{15}\frac{\text{e}^+}{\text{s}}}$$

REFLECT
Remember that the mass number is the total number of protons *and* neutrons. Sodium-22 has 11 protons and 11 neutrons, while neon-22 has 10 protons and 12 neutrons.

28.55

SET UP
The strong force acts over distance of approximately 1 fm. We're interested in the nucleus that has a diameter equal to ten times this, or with a radius of $r = 5.0$ fm. We can use $r = r_0 A^{\frac{1}{3}}$, where $r_0 = 1.2$ fm, to calculate the mass number of this nucleus, and then consult a periodic table (or Appendix C) to see which element has an atomic mass similar to this. As an approximation, we can treat the outer proton of the nucleus as being repelled by all of the inner protons located at the center of the nucleus and use Coulomb's law to calculate the magnitude of this repulsion.

SOLVE

Part a)

$$r = r_0 A^{\frac{1}{3}}$$

$$A = \left(\frac{r}{r_0}\right)^3 = \left(\frac{5.0 \text{ fm}}{1.2 \text{ fm}}\right)^3 = 72.3$$

Consulting the periodic table, germanium (Ge) has an atomic weight of 72.6, so a germanium nucleus would have a diameter equal to about 10 times the range of the strong force.

Part b)
Germanium has an atomic number of 32. We will treat the outer proton as being repelled by the 31 inner protons located at the center of the nucleus:

$$F = \frac{k(e)(31e)}{r^2} = \frac{\left(8.99 \times 10^9 \frac{\text{N} \cdot \text{m}^2}{\text{C}^2}\right)(31)(1.6 \times 10^{-19} \text{ C})^2}{(5.0 \times 10^{-15} \text{ m})^2} = \boxed{290 \text{ N}}$$

REFLECT
The strong force opposes this electrical repulsion and keeps the nucleus together.

28.59

SET UP
Muonic hydrogen is a hydrogen atom where the electron has been replaced by a muon. We can calculate the Bohr radius and ground-state ionization energy of muonic hydrogen by setting up a ratio with the values for regular hydrogen ($a_0 = 5.29 \times 10^{-11}$ m, $E_0 = 13.6$ eV). The algebraic expressions for the Bohr radius and the ground-state energy of hydrogen (also known as the Rydberg energy) are $\frac{\hbar^2}{mke^2}$ and $\frac{m(ke^2)^2}{2\hbar^2}$, respectively; the mass is either the

mass of the electron ($m_e = 0.5110$ MeV/c^2) or the mass of the muon ($m_\mu = 105.7$ MeV/c^2). The big difference in ionization energies arises from the difference in electrostatic attraction between the muon and the nucleus and the electron and the nucleus.

SOLVE

Part a)

$$\frac{a_0(\mu)}{a_0(e)} = \frac{\left(\dfrac{\hbar^2}{m_\mu ke^2}\right)}{\left(\dfrac{\hbar^2}{m_e ke^2}\right)} = \frac{m_e}{m_\mu} = \frac{\left(0.5110\dfrac{\text{MeV}}{c^2}\right)}{\left(105.7\dfrac{\text{MeV}}{c^2}\right)} = 0.00483$$

The Bohr radius in muonic hydrogen is $\boxed{4.83 \times 10^{-3} \text{ times smaller}}$ than the Bohr radius of ordinary hydrogen.

$$a_0(\mu) = (0.00483)a_0(e) = 0.00483(5.29 \times 10^{-11} \text{ m}) = \boxed{2.56 \times 10^{-13} \text{ m} = 2.56 \times 10^{-4} \text{ nm}}$$

Part b)

$$\frac{E_0(\mu)}{E_0(e)} = \frac{\left(\dfrac{m_\mu(ke^2)^2}{2\hbar^2}\right)}{\left(\dfrac{m_e(ke^2)^2}{2\hbar^2}\right)} = \frac{m_\mu}{m_e} = \frac{\left(105.7\dfrac{\text{MeV}}{c^2}\right)}{\left(0.5110\dfrac{\text{MeV}}{c^2}\right)} = 207$$

The minimum ionization energy of ground-state muonic hydrogen is $\boxed{207 \text{ times}}$ larger than the ionization energy of ordinary ground-state hydrogen.

$$E_0(\mu) = (207)E_0(e) = 207(13.6 \text{ eV}) = \boxed{2810 \text{ eV}}$$

Part c) The muon has the same charge as the electron but it is much closer to the nucleus than the electron in ordinary hydrogen. Therefore the muon feels a much stronger electrical attraction from the nucleus, so it takes much more energy to remove it from the atom than is required for the electron in ordinary hydrogen.

REFLECT
The muon is a lepton, just like the electron, so it is not affected by the strong force.

mass of the electron, $m_e = 0.5110$ MeV/c² for the mass of the muon ($m_\mu = 105.7$ MeV/c²). The big shift in ionization energies arises from the difference in electrostatic attraction between the muon and the nucleus and the electron and the nucleus.

SOLVE

Part (a)

The Bohr radius in muonic hydrogen is 2.56×10^{-13} m. Therefore, it is much smaller than ordinary hydrogen.

Part (b)

The ionization energy of the ground state in muonic hydrogen is about 207 times larger than the ionization energy of ordinary hydrogen in its ground state.

REFLECT